Physical Geography

Physical Geography

RICHARD S. PALM

University of Wisconsin—Eau Claire

Charles E. Merrill Publishing Company
A Bell & Howell Company
Columbus   Toronto   London   Sydney

Published by
Charles E. Merrill Publishing Company
A Bell & Howell Company
Columbus, Ohio 43216

This book was set in Times Roman and Helvetica Medium.
The cover was prepared by Will Chenoweth.

© 1978, by Bell & Howell Company. All rights reserved. No part of this book may be reproduced in any form, electronic or mechanical, including photocopy, recording, or any information storage and retrieval system, without permission in writing from the publisher.

Cover photograph by H. Armstrong Roberts.

Frame 13-3 native gold slide from *Minerals and Man,* by Cornelius S. Hurlbut, Jr. Published by Random House, Inc. in 1968. Reprinted by permission of Random House, Inc.

Frames 13-6, 13-12, 13-18, and 14-13, and Figure 14-6, from *Geology Today,* Revised Printing. Copyright © 1973, 1974 by Ziff-Davis Publishing Company. Reprinted by permission of CRM Books, a Division of Random House, Inc.

Library of Congress Catalog Card Number: 77-92951
International Standard Book Number: 0-675-08403-2
Printed in the United States of America
1 2 3 4 5 6 7 8 9 10/85 84 83 82 81 80 79 78

**To my wife Karen
and my children Christopher and Andrew**

*Without their patience, tolerance, and love,
this project would not have been completed.*

Preface

This physical geography program is designed to introduce students to the many aspects of the physical environment. The major goal of the program is to instill in students a greater appreciation and awareness of man's environment. With this program, students will gain an understanding of how physical geography relates to their lives. For example, the spatial pattern of population on the earth is partly a response to the advantages and problems associated with highly diversified physical environments. In addition, we react daily to changes in the physical envirionment: Will I need my raincoat today? What kind of fertilizer should I put on my lawn? Is June a good month to travel in Europe? To answer these basic questions, and many others that are more complex, one must have knowledge of the physical environment. The variability of landforms, the great diversity of vegetation and soils, and the daily and seasonal changes in the atmosphere from place to place are all fascinating and timely topics. The author hopes that this program will generate student desire for further knowledge of more advanced studies of the physical environment.

A multimedia approach was selected for this program because of its advantages for both students and instructor. Students benefit from the precise audio in concert with the full-color visuals. The textbook material reinforces the learning experience by its close correlation to the audio-visual portion and can therefore serve as either the primary or the review-study source. If the program is used for individualized instruction, the instructor will have more freedom to devote to students on a one-to-one basis or to lead small student discussion groups.

The program is divided into two broad categories: (1) weather and climate, and (2) landforms. Vegetation and soils follow the weather and climate modules because of their close relationship to the world patterns of moisture and temperature.

Module 1 analyzes the development of the Geographic Grid System and explains the causes for changing seasons on the earth. The Additional Material section analyzes time changes around the earth. Modules 2 through 6 examine the elements that make up the weather: the energy budget and temperature, atmospheric pressure and winds, moisture and precipitation, atmospheric circulation, and severe storms. In addition, these modules include explanations of heating and cooling degree days, the wind chill factor, temperature scales, clouds, and adiabatic lapse rates. Module 7 examines the hydrologic cycle and the mechanics of the local water budget. Module 8 summarizes the weather elements in world climate regions based on Köppen's climate classification.

Modules 9 through 12 explain the factors that cause soils and vegetation to vary from place to place. These modules also describe the world pattern of soils and vegetation by use of the most recent U.S. comprehensive soil classification system (7th Approximation System). Modules 13 through 19 introduce the student to rocks and minerals, the factors that shape the earth's surface, and the features formed by those factors. Vulcanism, diastrophism, and weathering, and erosion and deposition by running water, gravity, glacial ice, wind, and waves, are discussed in detail. The entire Module 14 is devoted to the relatively new topic of plate tectonics and its effects on the earth's surface features. Maps, a fundamental tool in geography, are introduced in the Appendix.

The author wishes to thank the many individuals who contributed to the program, both with their recommendations and with their slides and photographs. Special thanks go to the following colleagues at the University of Wisconsin—Eau Claire for their invaluable assistance: Professors Cahow, Foust, de Souza, Janke, Tanner, Vogeler, and Willis. The author is also greatly indebted to Maggie Shaffer for editing the program and making numerous suggestions for improvement.

To the Student

This program is designed to introduce you to physical geography using a variety of media. MEDIAPAK audiovisual presentations are integrated with this textbook and can be used throughout the course to enable you to master concepts of physical geography faster and more thoroughly than if you used a conventional text alone.

Format of the Text

This program is divided into nineteen separate units, or modules. Each module contains the following:

Introduction—A brief look at the material to be covered.

Learning Objectives—Statements of what concepts should be learned.

Key Terms—A list of difficult terms, and words that might be unfamiliar to you. When these terms are discussed in the text, they appear in **boldface** type.

Outline—An outline of the audiovisual presentation, which will be helpful for reference and review.

Exercises—Questions and problems designed to reinforce your learning.

Summary—A review of the audiovisual presentation; contains a number of the key visuals from the audiovisual materials for further study.

Questions—Problems designed to help you probe more deeply into the concepts presented and to relate these concepts to ideas presented in other modules.

Suggested Readings—A list of additional available sources of the topics covered.

Some modules contain Additional Material sections, which present material related to the topics presented in those modules. The textbook also features a Glossary, which contains definitions of most Key Terms and other important concepts, and an Appendix, entitled "Maps."

Procedures

Before going to the audiovisual presentation for a given module, you should read the corresponding Introduction and study the Learning Objectives and Key Terms. Then, select the MEDIAPAK component(s) for that module. You may stop your audiovisual equipment if you need more time on a particular visual frame. Also, if you do not understand a concept, you may rewind the equipment and review the information. After completing the audiovisual portion of the module, return to the textbook and answer the Exercises. You should read the Summary for review and further study of the material covered in the audiovisual sequence.

Contents

Module 1

Location and Seasons   1

Module 2

The Energy Balance and Temperature   15

Module 3

Atmospheric Pressure and Winds   35

Module 4

Atmospheric Moisture and Precipitation   53

Module 5

Atmospheric Circulation   73

Module 6

Severe Storms   91

Module 7

Hydrologic Cycle and Local Water Budget   107

Module 8

World Climatic Regions   119

Module 9
Soil Properties and Processes 145

Module 10
World Soil Patterns 159

Module 11
Vegetation Controls 171

Module 12
World Vegetation Patterns 181

Module 13
Landform Composition 199

Module 14
The Shifting Continents 211

Module 15
Vulcanism and Diastrophism 227

Module 16
Weathering and Mass Movement 245

Module 17
Landforms from Running Water 257

Module 18
Landforms from Glacial Action 279

Module 19
Landforms from Wind and Waves 299

Appendix
Maps 317

Acknowledgments 331

Glossary 334

Index 346

Physical
Geography

Module 1

Location and Seasons

INTRODUCTION

Although many people believe that Columbus was the first person to prove that the earth is round, he was neither the first to hypothesize that concept nor the first to prove it. As early as the fifth century B.C., the Greek mathematician Pythagoras proposed that the earth was round. By 200 B.C., Eratosthenes, another Greek mathematician, had calculated the circumference of the earth based on the size of shadows cast by the sun at Alexandria and Syene; today we know that his calculations were only 200 miles short of the actual circumference. A short time later (about 150 B.C.), Poseidonius, also a Greek mathematician, calculated the circumference of the earth at only three-fourths of the actual distance. Still later, in his writing entitled "Geographia," Greek astronomer and geographer Ptolemy used the philosophy of a round earth to make map projections with latitude and longitude in degrees. Unfortunately, however, he accepted the calculations of Poseidonius. Therefore, although later voyagers such as Columbus already had strong evidence that the earth was round, they believed the earth to be much smaller than it actually is. For example, after landing in the Bahamas, Columbus believed that he had already sailed past Japan by one week. In fact, he was so sure that he was in Asia that when he reached the coast of Central America, he identified it as Indochina. Magellan's trip around the world a few years later confirmed that the distance around the earth was much greater than expected and that the world was indeed round.

Newton in 1687 proposed that the earth was an oblate elliptical shape (that is, flattened at the poles); and geodesists, or people who study the shape of the earth, held this belief for many years. Modern geodesists, however, are questioning this belief and have proposed a new shape called the *geoid,* a pearlike shape with an irregular surface.

The satellite LAGEOS (LAser GEOdynamic Satellite) has been launched to seek further information about the earth's shape. For all practical purposes, however, such as for mapping, surveying, and determining time, location, and distance, scientists will continue to treat the earth as a spherical surface. Therefore, it is important that any student of the physical earth understand the basic concepts of both location on and distance around a spherical world, as well as shape and size of the earth, all of which are fundamental to an understanding of seasons and of night and day.

OBJECTIVES

By the end of this module, you should be able to do the following:

1. Describe the earth's basic shape and the distribution of land and water on the earth's surface.
2. Explain the relationship of meridians and parallels to longitude and latitude, respectively.
3. Use latitude and longitude to locate places on the earth.
4. Explain the relationship of (a) the earth's revolution, (b) parallelism, and (c) inclination of the earth's axis to the seasons.
5. Explain the significance of the Tropics of Cancer and Capricorn, and the Arctic and Antarctic circles.
6. Describe the distinction between the vertical ray of the sun and its more oblique rays.
7. Identify the dates for the solstices and equinoxes and explain the conditions that make the solstices and equinoxes outstanding from other dates.

KEY TERMS

oblate ellipsoid	rotation
coordinate	plane of the ecliptic
grid system	circle of illumination
meridian	perihelion
parallel	aphelion
degree	inclination
latitude	parallelism
longitude	solstice
prime meridian	equinox
revolution	vertical ray

Now you are ready to begin the audiovisual portion of this module. Select the MEDIAPAK 1 component(s) and proceed. Following is a topical outline of the audiovisual sequence. You will find this outline helpful for reference and review. After completing MEDIAPAK 1, return to this book to perform the exercises.

OUTLINE

Earth data
Earth shape
Land and water

Location
Geographic grid system
Latitude and longitude

Seasons
Rotation and revolution
Inclination and parallelism
Solstices and equinoxes
Vertical rays vs. oblique rays

Exercises

1. The earth has been called the *water planet*. What percentage of the earth's surface is water?
   - **A.** 10
   - **B.** 30
   - **C.** 50
   - **D.** 70
   - **E.** 90

2. Parallels run in an east-west direction.
   - **A.** True
   - **B.** False

3. Longitude is used to determine the location of places
   _____________________ of the _____________________.
   - **A.** north and south; equator
   - **B.** east and west; equator
   - **C.** north and south; prime meridian
   - **D.** east and west; prime meridian

4. The earth's axis is inclined from a vertical position by _____ degrees.
   - **A.** 12
   - **B.** 23½
   - **C.** 90
   - **D.** 66½
   - **E.** 45

5. On the June solstice, the _____________ Pole is inclined toward the sun.
   - **A.** North
   - **B.** South

6. On the December solstice, the vertical ray of the sun is directly overhead at the
   _____________________.
   - **A.** Tropic of Cancer
   - **B.** Tropic of Capricorn
   - **C.** Arctic Circle
   - **D.** Antarctic Circle

7. Some latitudes on the earth never receive vertical rays during a year, while some latitudes receive vertical rays twice each year.
   - **A.** True
   - **B.** False

8. The potential heating effect of the sun's rays is greatest with oblique rays because they cover a greater total surface area than do vertical rays.
   - **A.** True
   - **B.** False

9. The dates when the vertical ray of the sun strikes at the equator are called the
   _______________. Their calendar dates are
   _______________________________.

   A. solstices; Dec. 21 and June 21
   B. solstices; March 21 and Sept. 21
   C. equinoxes; Dec. 21 and June 21
   D. equinoxes; March 21 and Sept. 21

10. The latitude of the Arctic and Antarctic circles is ______ degrees.
    A. 23½
    B. 90
    C. 75½
    D. 45
    E. 66½

Summary

Earth Data

Geodesic research has shown that the earth is not a true sphere. More precisely, the earth has the form of an **oblate ellipsoid;** that is, the earth bulges at the middle and is slightly flattened at the top and bottom. The earth's radius from pole to pole is a distance of 7900 miles (12,714 km), while its radius at the equator is 7929 miles (12,757 km). Clearly, the earth is slightly broader across the middle. However, for all practical purposes, such as for calculating distances and time changes, the earth is considered to function as a sphere. Working with a true sphere simplifies mathematical calculations, and since the variation of the earth's shape from a sphere is so slight, very little accuracy is lost by treating it as such.

To geographers, the outer surface, rather than the shape, of the earth is of most importance, for they analyze the relationship between people and their environment. The earth's outer surface is much smoother than most people realize. A variation in elevation of only 12 miles between the highest mountain peak and the deepest ocean trench is a very minor variation in elevation in relationship to the earth's total size. Indeed, the average elevation of land surfaces is only about one-half mile above sea level.

Not only is the surface of the earth very smooth, but also it varies considerably in the distribution of its land and water surfaces. About 70% of the earth's surface is occupied by the world's oceans. Of the 30% of the world's surface covered by land, slightly more than two-thirds lies in the Northern Hemisphere.

Since geographers are interested in the relationship between people and their environment, they are also quite concerned about the world's ever-increasing population and its effect on the earth's surface. People are not evenly distributed over the earth's land surfaces but instead are very selective of their habitats. A glance at a world population density map in any atlas reveals the pockets of extremely dense population and the large sections of land surface nearly devoid of people. Two-thirds of the people on earth live at elevations below 600 feet (1968 m). However, increased population pressures may eventually force large numbers of people to seek less desirable locations with enviromental problems that will require specialized adaptations to be able to support a dense population.

Location

The cornerstone of geographic inquiry is place location. Without the ability to locate places, people would be unable to communicate or move even moderate distances around the world.

To locate a place accurately requires the use of at least two reference lines. On the earth's surface, a series of imaginary reference lines, called grid lines or **coordinates,** criss-crosses the surface to form a **grid system** from which measurements of distance can be made to determine different locations. This grid system, known as the *Geographic Grid System,* is applicable to all places on the earth.

The series of reference lines that run north to south, extending from pole to pole in the Geographic Grid System, is composed of **meridians.** Meridians are used to determine place locations east or west of the **prime meridian.** The second series of reference lines in the Geographic Grid System is composed of **parallels.** Parallels run in an east-west direction, and their name derives from the fact that they lie parallel to the equator. Parallels are used to determine locations north or south of the equator. Together, meridians and parallels comprise a grid system that is used to locate places on the earth.

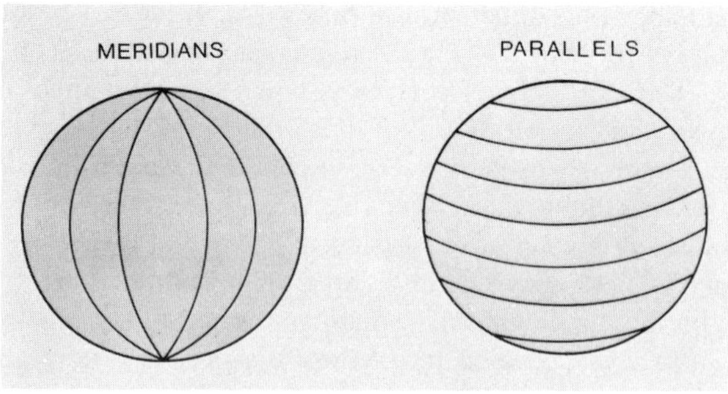

Figure 1–1

Due to the curvature of the earth's surface, meridians and parallels are not straight lines but rather arcs on a sphere. An arc is a fraction of a circle, and fractions of a circle are measured in **degrees.** Therefore, locations and distances along an arc are measured in degrees from a given zero point.

Parallels are used to measure **latitude,** the degree of arc measured north and south of the equator. The equator divides the earth into northern and southern halves or hemispheres. It is labeled zero degrees (0°) latitude, with each successive degree of latitude labeled in a north or a south direction to either pole. Each pole is exactly 90° from the equator. The latitude of a location is determined by the parallel that passes through that location. For example, if the 45° parallel north of the equator passes through the place in question, the latitude of that place is 45° north. Remember, it is essential to label degrees of latitude as either north or south of the equator.

Meridians are utilized to measure **longitude,** the degree of arc measured east and west of the prime meridian. The prime meridian is labeled as zero degrees (0°) longitude, with each successive degree of longitude labeled in an east or a west direction to a meridian halfway around the earth at 180°. The longitude of a location is determined by the meridian that passes through that place. For example, if the 90° meridian west of the prime meridian passes through the place in question, the longitude of that location is 90° west. Once again, it is necessary to indicate whether the longitude is east or west because the prime meridian and 180th meridian divide the earth into eastern and western halves or hemispheres.

To locate a place, one must state both the degrees latitude and the degrees longitude. For more precise measures of location, whole degrees can be subdivided into smaller units such as minutes and seconds. Places labeled in whole degrees will be accurate within 69 miles (111 km). Subdividing degrees into minutes increases the accuracy of the location to 1.2 miles (2 km), and subdividing minutes into seconds reduces the margin of error to about 100 feet (31 m). For locations more precise than 100 feet, local survey systems would apply. Within the limits indicated, it is possible to locate any place on earth using the Geographic Grid System.

Seasons

The earth simultaneously goes through two motions. The earth orbits or revolves around the sun once every 365¼ days, or one year; this motion is termed **revolution**. The earth's orbit about the sun is elliptical. The earth also turns on its axis once every twenty-four hours; this motion is called **rotation.**

The average distance between the earth and the sun is 93 million miles. However, due to the ellipticity of its orbit, the earth is as close as 91½ million miles at a time in its orbit called **perihelion** (about January 3), and as far away as 94½ million miles at a time called **aphelion** (about July 4). This change in distance between the earth and sun does cause some variation in the amount of sun's energy received on earth, but this variation does not cause the seasons. This fact becomes apparent when we remember that at perihelion, when the earth should receive a greater proportion of sun's energy, it is January and the coldest time of the year in the Northern Hemisphere. At the same time, while the Northern Hemisphere experiences its coldest period of the year, the Southern Hemisphere experiences its warmest period. Instead, two factors other than perihelion and aphelion generally control the change of seasons.

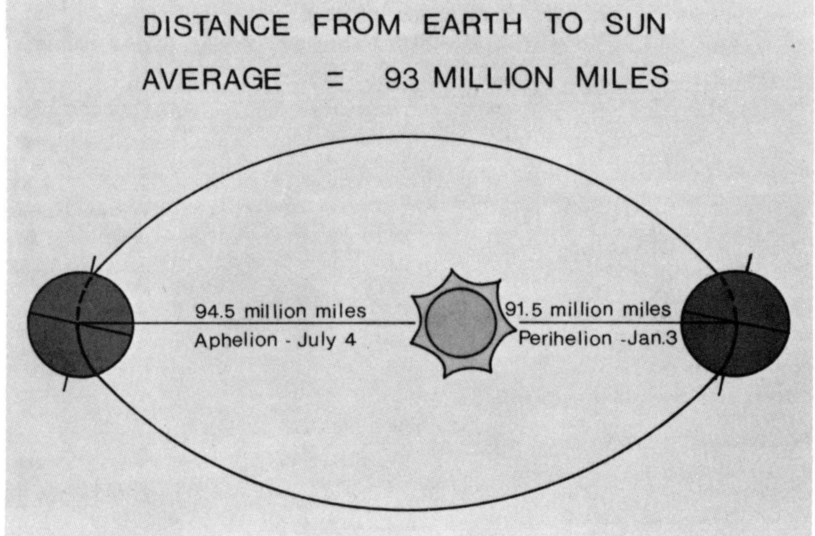

Figure 1–2

One factor is the **inclination** of the earth's axis. The earth's axis, which extends between the North and South poles, does not lie at a right angle to the earth's orbital plane. Instead, the axis is inclined at an angle of 23½° from the vertical. The orbital plane, or **plane of the ecliptic,** is the imaginary plane described by the orbital path of the earth around the sun. The second factor controlling the seasons is the fact that as the earth orbits the sun, the earth's axis is always inclined in the same direction in reference to outer space. In other words, the earth's axis is parallel to itself throughout its orbital path. This condition of constant orientation is called **parallelism.** Figures 1–3 and 1–4 will help to explain the effects of an inclined axis and parallelism on the receipt of sun energy on the earth, which is, in turn, ultimately responsible for seasonal change.

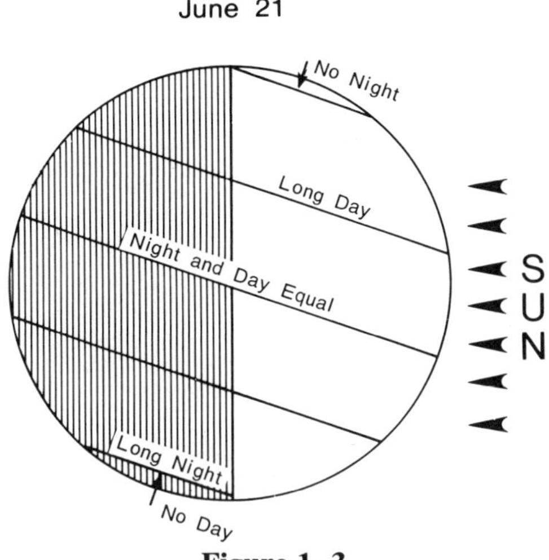

Figure 1–3

At any given time, one-half of the earth is in daylight and the other half in darkness. The boundary between day and night is referred to as the **circle of illumination.** Note that in June (Figure 1–3), the circle of illumination does not evenly bisect the latitudes on earth. At 50° north latitude, more of the latitude line is in daylight than in darkness. At the same time, just the opposite is true at 50° south latitude; that is, a greater part of the latitude line is in darkness. In fact, in June, all latitudes north of the equator experience more hours of daylight during a 24-hour period than they do hours of darkness; in the Southern Hemisphere just the reverse is true. The reason that there are more daylight hours in the Northern Hemisphere in June is that the North Pole is inclined toward the sun at that time of year due to the inclination of the earth's axis. Consequently, the Northern Hemisphere receives more direct solar energy than does the Southern Hemisphere. This greater energy causes greater heating in the Northern Hemisphere and explains the summer season in that hemisphere. Since the Southern Hemisphere is inclined away from the sun in June, places in the Southern Hemisphere are in the middle of their winter season.

In December (Figure 1–4), the earth has orbited to a position on the opposite side of the sun. Due to parallelism, however, the axis is still inclined in the same direction. As a result, in December the South Pole is inclined toward the sun, and Southern Hemisphere latitudes receive the most direct solar energy. Thus, December is a summer month in the Southern Hemisphere. Simultaneously, in the Northern Hemisphere just the reverse is true.

Heating of the earth's surface by solar energy is based on the simple fact that the greater the concentration of energy per unit area, the greater is the potential heat. If the earth's surface were a flat plane, all places on earth

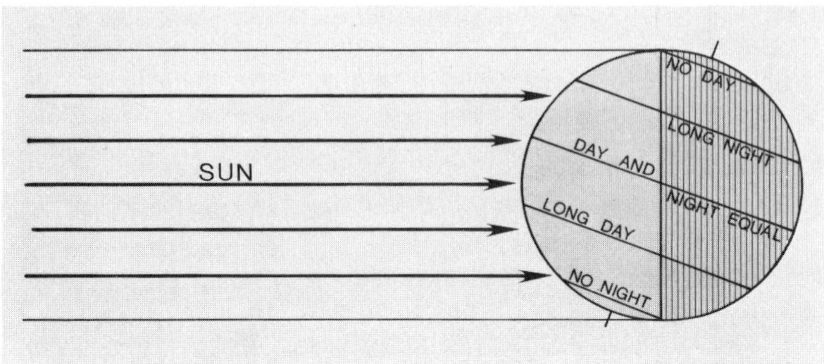

Figure 1–4

would receive the sun's energy at a direct 90° angle, or by a **vertical ray.** Because the earth's surface is curved, however, only one latitude on earth on any given day receives the vertical ray of the sun; all other latitudes receive solar energy at angles less than 90° or by oblique rays. The more oblique the ray of energy, the less is the amount of energy absorbed per unit area, and the less is the potential heating effect.

If the earth's axis were not inclined but instead were at right angles to the orbital plane, the vertical rays of the sun would always strike the earth's surface directly at the equator. In that case, both the Northern and the Southern hemispheres would receive equal amounts of energy all year and there would be no seasons. But because the earth's axis is inclined, and because of parallelism, the vertical ray of the sun strikes the earth's surface at latitudes ranging between 23½° north to 23½° south during the year. On June 21 **(June solstice),** the vertical ray of the sun strikes at 23½° north latitude (Tropic of Cancer), the northernmost latitude to which the vertical ray extends each year. As the earth orbits to the opposite side of the sun during the next six months, the vertical ray of the sun migrates southward each day. On December 21 **(December solstice),** the vertical ray strikes at the southernmost latitude (23½° south), called the Tropic of Capricorn.

From December 21 to June 21, the vertical ray begins its migration northward. Twice each year, once as the vertical ray shifts south and again as it shifts north, the vertical ray strikes exactly on the equator. On those dates (March 21 and September 21), the circle of illumination passes through both poles. Consequently, all locations on earth experience 12 hours of daylight and 12 hours of darkness in a 24-hour period. Because the lengths of day and night are equal, those dates are referred to as the **equinoxes.**

On the solstices, poleward of latitude 66½° in each hemisphere, the earth is enveloped in either 24 hours of daylight or 24 hours of darkness. On June 21, poleward of 66½° north (Arctic Circle), all places receive 24 hours of daylight. On December 21, places poleward of 66½° south (Antarctic Circle) receive 24 hours of daylight. Of course, while one polar region is experiencing continuous daylight, the other polar region is in continuous darkness.

Additional Material

Time Changes

If you have ever traveled by airplane from one continent to another, you undoubtedly experienced time changes as you traveled east or west. In some cases the time changes may have been great enough to temporarily affect you physically, and you may have felt very tired for several days until your body adjusted to the new time zone. (This malady is commonly referred to as *jet lag*.) But why does time change from place to place?

The entire concept of time is based on the relationship between the earth's rotation and the sun. As we have seen, the earth makes one complete rotation every 24 hours and is, for all practical purposes, a sphere that contains 360°. Since the earth turns 360° every 24 hours, it turns 15° every hour. Consequently, longitudes 15° apart are one hour apart in time.

If people were to set their watches on sun time, each person would be walking around with a different time and would have to adjust his watch as he moved from one meridian to the next. To avoid this inconvenience, countries established standard time zones numbered from east to west beginning at the international date line, which coincides roughly with the 180° meridian. These times zones stretch from pole to pole, and each has uniform time throughout. There are 24 time zones, each 15° wide, with the time in each zone based on the time at the central meridian that lies in the middle of the zone. For example, the central standard time zone (CST) in the United States is based on the time at the 90° meridian.

When viewed from outer space, the earth rotates from west to east. In other words, the sun crosses the sky, or time progresses, from east to west. New days begin at the international date line, which generally coincides with the 180° meridian. The new day progresses around the earth in a westerly direction, advancing first through the time zones in the Eastern Hemisphere and later through the time zones of the Western Hemisphere, ending the day when it returns to the international date line. Therefore, in all places east of you, to the international date line, that are not a part of your time zone, you would see the new day before you; and these places would have later times than your time. In all places west of you, to the international date line, that are not included in your time zone, you would see the new day after you; and these places would have times earlier than your time.

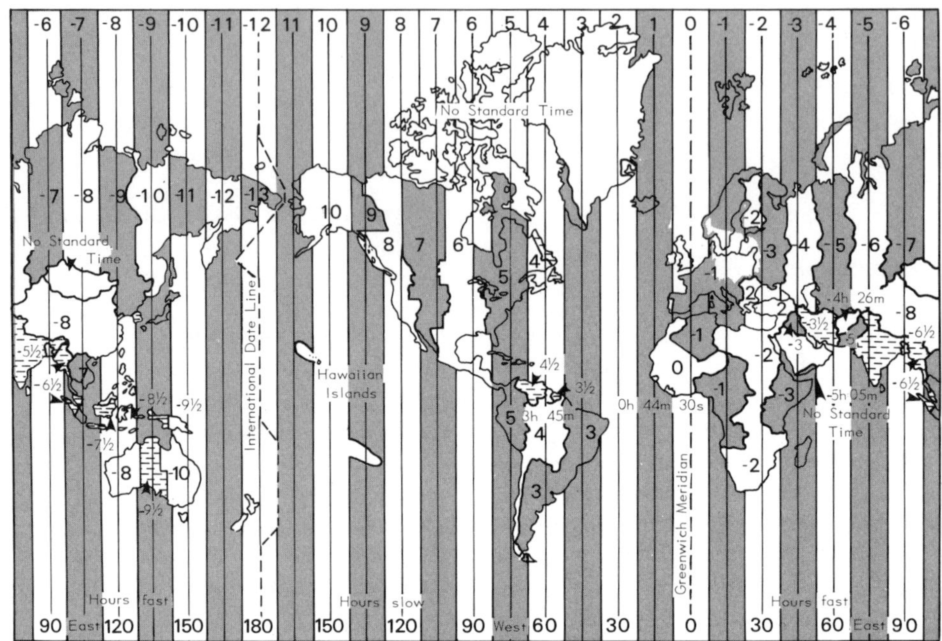

Figure 1–5

In the United States, for example, there are four standard time zones: the eastern standard time zone (EST), the central standard time zone (CST), the mountain standard time zone (MST), and the Pacific standard time zone (PST). For each time zone that you cross traveling west, you gain an hour. For example, if it is 1:00 P.M. EST in New York City, it will be four hours earlier in San Francisco, or 10:00 A.M. PST. Each time zone that you cross while traveling east results in the loss of an hour. New York City (73° west) and Hamburg, Germany (10° east), are six standard time zones apart. Therefore, if it is 1:00 P.M. in New York City, it will be six hours later in Hamburg, or 7:00 P.M.

Theoretically, it should be possible to continue traveling west and constantly gain time. It is possible, in fact, to gain as many as 23 hours, assuming that you started at the first standard time zone. It is not possible to gain more than 23 hours, however, because once you reach the 180th meridian, you are at the international date line. When you cross the international date line traveling west, you move from the Western Hemisphere into the Eastern Hemisphere. As a result, based on the standardization of time, you lose 24 hours, or one day. For example, let us suppose that you are in the Western Hemisphere near the international date line, and the time is 5:00 P.M. and the day is Thursday. You then cross the international date line into the Eastern Hemisphere. The time is still 5:00 P.M., but the day has changed to Friday.

(United States Navy vessels situated in the Pacific Ocean near the international date line have been known to deliberately cross the date line on occasion to gain a little extra liberty time. If it is Saturday in the Western Hemisphere, when the sailors cross into the Eastern Hemisphere, the day changes to Sunday, the sailors' day off. Then, once Monday begins in the Eastern Hemisphere, the ship crosses back into the Western Hemisphere, where the day is still Sunday. Thus, the sailors are able to get two days off instead of only one day.)

The margins of the time zones are not straight; rather, they have been adjusted for convenience. In the United States, for example, state boundaries are often selected as a time-zone boundary in place of a meridian to avoid having the same state in two different time zones. Even the international date line is sometimes altered, as in the Bering Strait, so that it passes between Asia and North America and around several Pacific islands.

Questions

1. How do the Northern and Southern hemispheres contrast in terms of the proportion of land and water surface on each?

2. Why are the reference lines in the Geographic Grid System arcs? Why are the reference lines labeled in degrees?

3. How accurate can you be by using whole degrees as the unit for specifying location? How accurate can you be by using minutes or seconds of a degree?

4. What is the relationship of parallels and meridians to latitude and longitude, respectively?

5. What would be the effect of increasing the speed of the earth's rotation?

6. What do you think would happen to the seasons if the earth's axis were inclined 45° instead of 23½°?

7. Would you expect any significant changes in the seasons if the earth's revolution were spherical instead of elliptical?

8. How do you explain the fact that the seasons on earth are just the reverse in the Northern and Southern hemispheres?

9. Explain the change in latitudinal position of the sun's vertical ray from December 21 to June 21.

10. Why are the most direct rays (vertical rays) from the sun the most effective in terms of heating?

11. Why do all latitudes on earth experience 12 hours of day and 12 hours of night on the equinoxes?

Suggested Readings

GREENWOOD, D. 1964. *Mapping*. Chicago: University of Chicago Press. Chapter 1.

KING-HÉLÉ, D. 1967. The shape of the earth. *Scientific American,* October 1967, pages 67–76. Reprint 873. San Franciso: W. H. Freeman.

LEHR, P.; BURNETT, R. W.; and ZIM, H. S. 1975. *Weather*. New York: Golden Press. Pages 48–52.

OLIVER, J. E. 1977. *Perspectives on applied physical geography*. North Scituate, MA: Duxbury Press. Chapter 1.

ROBINSON, A. H. 1966. *Elements of cartography*. 2d ed. New York: John Wiley & Sons. Chapter 2.

SMYLIE, D. E., and MANSINHA, L. 1971. The rotation of the earth. *Scientific American,* December 1971, pages 80–88. Reprint 897. San Francisco: W. H. Freeman.

Module
2

The Energy Balance and Temperature

INTRODUCTION

We all know that temperature varies both on a daily basis and on a seasonal basis. We also know that generally it is the relationship between the earth and the sun that causes these fluctuations. But exactly what causes the basic fluctuations of temperature on the earth? Why is it important that we understand these changes?

The climates around the world have changed considerably in the past, which indicates that the relationship between the earth and the sun is not a constant condition. By analyzing fossils, pollen, tree rings, and human artifacts, scientists have been able to piece together enough knowledge to make at least general statements about past climates. We know, for example, that there were at least four major periods of glaciation on earth, at which times the overall world temperature was several degrees Celsius cooler than it is today. Between the time of the last major glaciation, about 10,000 years ago, until today, there have been many periods when the climate has been slightly warmer or slightly cooler than it is today. Recent evidence suggests that we have been in a period of cooling on earth since the early 1940s. Whether or not this trend will continue into a full-fledged "Ice Age" is still a matter of conjecture. At this time we are unable to make such long-range predictions; but if we were, think what a tremendous advantage it would be to know within a reasonably advanced period of time the future climates of all places on the earth. The implications of such knowledge for agriculture, industry, business, and recreation are overwhelming.

We are most concerned today with being able to meet the increasing demands for energy on the earth. The fossil fuels which we have been utilizing in exorbitant amounts for years, and which ultimately came from the sun, now appear to be available in finite amounts. Thus, in the search for substitute energy forms, considerable attention has been focused on tapping the sun's energy directly via various solar-energy–absorbing systems, such as solar-heated homes and solar furnaces. Although it is neither economically nor technically feasible to shift to large-scale solar energy production at this time, it is not far into the future when solar energy and other forms of energy, such as geothermal and nuclear, will be required regardless of their cost. It is important, therefore, that we understand the atmospheric changes that might change the amount of solar energy received on the earth.

The sun provides the earth-atmosphere system with 99.9% of its energy, but this figure does fluctuate slightly. Changes in the amount of solar energy received on earth are often due to minor fluctuations in the sun's output, over which we have no control. The changes that we have created in the atmosphere, however, can have as much an effect on the amount of solar energy that we receive as do the sun's natural fluctuations. Environmentalists have voiced concern over a wide variety of issues that affect our atmosphere: air pollution from automobiles and industrial activities; depletion of ozone in the atmosphere by the release of fluorocarbons used to propel aerosols from spray cans; and the new supersonic transport (SST), which flies at high altitudes and is known to release nitrogen oxides that can deplete the protective ozone in the atmosphere. The depletion of ozone in the atmosphere by just 5% could sharply increase the incidence of skin cancer in humans as well as disrupt the growth of major agricultural products and curtail the growth of ocean plankton, an essential part of the ocean's food chain. It should be apparent that the relationship between the sun and the earth is critical for all forms of life.

OBJECTIVES

By the end of the module, you should be able to do the following:

1. List and describe the layers of the atmosphere, and explain how temperature varies with each layer.
2. Explain the factors that are responsible for reflecting and scattering incoming solar radiation.
3. Compare the greenhouse effect with the conversion of solar radiation to long-wave radiation at the earth's surface.
4. Explain the outgo of energy from the earth's surface.
5. Explain how heat is transferred via radiation, conduction, and convection.
6. Describe the factors that are responsible for variations in total insolation.
7. Explain the pattern of temperature variation in a 24-hour period.
8. Explain how temperatures vary with latitude and altitude.
9. Explain the derivation of the various temperature statistics recorded routinely by local weather stations.

KEY TERMS

climate	conduction
weather	convection
solar radiation	latent heat of vaporization
solar constant	temperature
selective scattering	mean temperature
albedo	lapse rate
insolation	temperature inversion
long-wave radiation	isotherm
counter radiation	

Now you are ready to begin the audiovisual portion of this module. Select the MEDIAPAK 2 component(s) and proceed. Following is a topical outline of the audiovisual sequence. You will find this outline helpful for reference and review. After completing MEDIAPAK 2, return to this book to perform the exercises.

OUTLINE

Weather and climate

Layers of the atmosphere

Solar energy
Solar radiation spectrum and solar constant

Energy balance
Incoming solar radiation
Insolation and outgoing long-wave radiation
World pattern of energy
Causes for variations

Temperature
Daily pattern of temperature
Causes for variations
Statistics
World pattern of temperature

Exercises

1. The main layers of the atmosphere from the surface to the top of the atmosphere are the
   - **A.** troposphere, stratosphere, mesosphere, and thermosphere.
   - **B.** stratosphere, mesosphere, thermosphere, and troposphere.
   - **C.** troposphere, mesosphere, stratosphere, and thermosphere.
   - **D.** thermosphere, mesosphere, troposphere, and stratosphere.

2. Temperatures in the atmosphere reach a maximum in the
   - **A.** mesosphere.
   - **B.** troposphere.
   - **C.** thermosphere.
   - **D.** stratosphere.

3. By volume, the atmosphere is composed of _______ percent nitrogen, _______ percent oxygen, and _______ percent rare gases.
   - **A.** 21; 78; 1
   - **B.** 78; 21; 1
   - **C.** 45; 44; 1
   - **D.** 1; 21; 78
   - **E.** None of the above

4. The following factors are responsible for reducing the total solar energy that reaches the earth's surface. Match the average percentage to each factor.

Choice		Factors	Percentage	
_______	**1.**	Dust particles	**A.**	3
_______	**2.**	Gas molecules	**B.**	5
_______	**3.**	Clouds	**C.**	6
_______	**4.**	Water vapor in clouds	**D.**	15
_______	**5.**	Surface albedo	**E.**	21

5. Generally speaking, the higher the latitude, the ___________ will be the total insolation.
   - **A.** greater
   - **B.** less

6. Long-wave radiation from the earth's surface is more readily absorbed as heat in the atmosphere than is direct incoming solar radiation.
   - **A.** True
   - **B.** False

7. Via turbulent heat exchange, warm air heated by conduction at the earth's surface is transported upward by convection as cool air replaces the warm air at the surface.
   - **A.** True
   - **B.** False

8. Locations on or near large water bodies generally experience a __________ range of temperature than land locations.
   A. greater
   B. smaller

9. On a daily average basis, temperatures tend to be at a maximum
   A. in early morning.
   B. in mid-morning.
   C. at noon.
   D. in mid-afternoon.
   E. in early evening.

10. If the temperature at sea level (zero feet elevation) is 75°F, what is the temperature at 10,000 feet, assuming that the normal lapse rate applies?
    A. 110°F
    B. 69°F
    C. 40°F
    D. 15°F
    E. 10°F

11. The following diagram shows the pattern of the 50°F isotherm as it crosses from a water surface to a land surface and back over water again. During what season would this pattern be most likely?
    A. summer
    B. winter

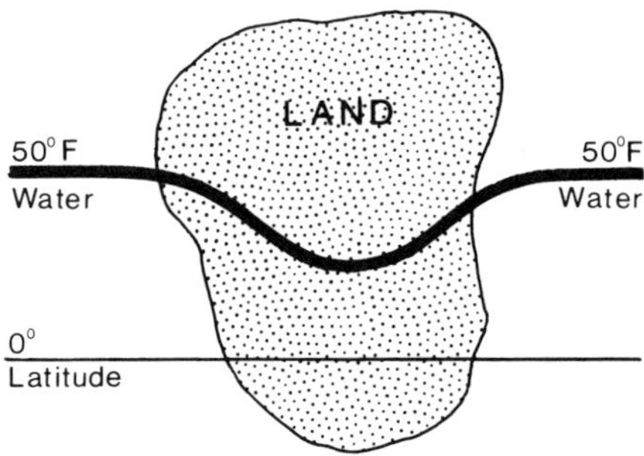

12. The following temperature data are monthly mean temperatures for each month during the year:

Month	J	F	M	A	M	J	J	A	S	O	N	D
Temp. (0°F)	15	21	32	40	52	66	72	70	68	59	42	29

From these data, the annual range of temperature is _______ degrees.
    A. 37°
    B. 40°
    C. 15°
    D. 57°
    E. 87°

Summary

One can rarely get through a day without hearing some description of the **weather.** The conditions of the atmosphere are described for us periodically on radio and television in terms of current temperature, precipitation amount, barometric pressure, wind speed and direction, relative humidity, and dew point. Over a long period of time, average day-to-day weather conditions in an area form a general pattern of weather that describes that area's **climate.**

The Atmosphere

The atmosphere is comprised of a series of zones that are differentiated from one another largely by the variation of temperature within each zone. As shown in Figure 2–1, the four zones are, in ascending order from the earth's surface, the troposphere, the stratosphere, the mesosphere, and the thermosphere. We will discuss some of the characteristics of each of these four major zones.

The *troposphere* is of the greatest direct importance to us because all of the weather in the atmosphere forms in this layer. The atmosphere is densest in this zone because of the pull of gravity, and within this zone the mass of the air decreases with increasing distance from the earth's surface. Also, temperature, as well as mass, normally declines with increased elevation above the earth's surface in this zone.

In the *stratosphere,* temperature is relatively constant in the lower part of the zone but increases significantly in the middle layer, termed the *ozone* layer. The ozone layer has received considerable attention in recent years as concern rises over the possible destruction of this layer by the increased quantity of fluorocarbons that we release into the air by our use of aerosol cans. Ozone in the atmosphere is essential to us, for it absorbs most of the ultraviolet rays from the sun—rays that could be harmful in large amounts. Above the ozone layer, temperatures drop; and at the stratopause, which separates the stratosphere from the mesosphere, temperatures reach values comparable to those near the earth's surface.

In the uppermost layer of the atmosphere, called the *ionosphere,* two subzones are identified: the mesosphere and the thermosphere. Temperature declines steadily as altitude increases in the *mesosphere,* for the rarefied atmosphere at this elevation is not capable of absorbing much solar energy. Consequently, at the top of the mesosphere, at the mesopause, temperatures may reach minus one hundred degrees Celsius ($-100°C$, or $-158°F$). The *thermosphere* lies at the very top of the atmosphere. Oxygen molecules in this layer absorb enough solar energy to push temperatures to several thousand degrees Celsius. More important to us than the high temperatures in the

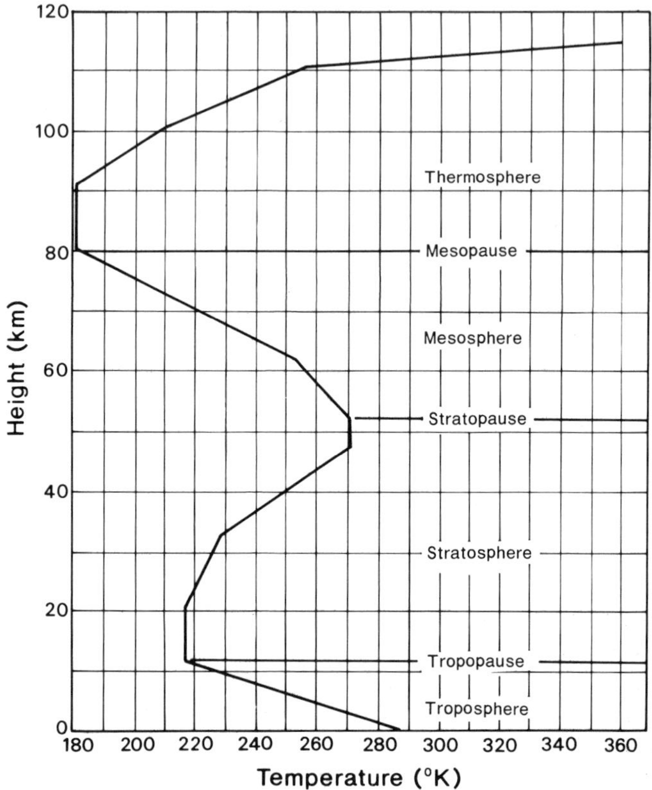

Figure 2–1

thermosphere is the ability of ions in this layer to reflect certain radio waves that are used for long-distance communications.

By volume, the atmosphere is composed of 78% nitrogen and 21% oxygen, with the remaining 1% various rare gases. (See Table 2–1.) Most of the gases are uniformly distributed through the atmosphere, although ozone is concentrated in the stratosphere and carbon dioxide and water vapor are concentrated near the earth's surface.

The Energy Balance

The driving mechanism for life on the earth is the sun, which provides 99.9% of our total energy. Like all other warm objects, the sun emits radiant energy, called **solar radiation.** Although the surface temperature of the sun fluctuates slightly with time, the amount of solar radiation given off to outer space is

Table 2–1

Gas	Percent of Atmosphere
Nitrogen	78.0
Oxygen	21.0
Rare Gases	
Argon	0.93
Carbon dioxide	0.03 to 0.1
Neon	0.018
Helium	0.005
Krypton	0.001
Hydrogen	0.005
Methane	0.002
Nitrous oxide	0.005

relatively constant. The amount of solar radiation that strikes the outer margin of the atmosphere is also nearly constant and, in fact, is referred to as the **solar constant** (equal to 1.95 calories/cm^2/minute). The earth-atmosphere system must return an amount of energy equal to the solar constant back to outer space. If the earth-atmosphere system absorbed more energy than it released, or released more than it absorbed, the overall temperature on the earth would rise or drop, respectively. But since the average temperature on the earth is nearly constant, an energy balance must exist within the system between the intake and outgo of energy.

Solar radiation is a collection or spectrum of energy waves of varying length that travel at uniform velocity through outer space. The solar radiation spectrum is shown in Figure 2–2. Only the visible light waves of the spectrum can be seen by the human eye. Sunshine, therefore, represents only a portion of the total solar radiation that reaches the earth's surface.

An energy balance can be compared to the mathematical balance of two sides of a ledger. (See Figure 2–3.) On one side of the ledger is incoming solar radiation, which enters the earth-atmosphere system and begins to penetrate toward the earth's surface. Portions of the incoming solar radiation never reach the surface, however, for some is *reflected* (or *scattered*) or *absorbed*. Minute dust particles in the upper atmosphere *scatter* the shortest waves of the visible light portion of the solar radiation spectrum (the blue portion of the color spectrum). Hence, the sky normally appears bluish in color. This process is termed **selective scattering** and is responsible for reducing total incoming solar radiation by 5%.

Gas molecules also intercept an average of 15% of incoming solar radiation, but by **absorption.** For example, oxygen molecules in the stratosphere combine photochemically with the ultraviolet rays of the spectrum to produce

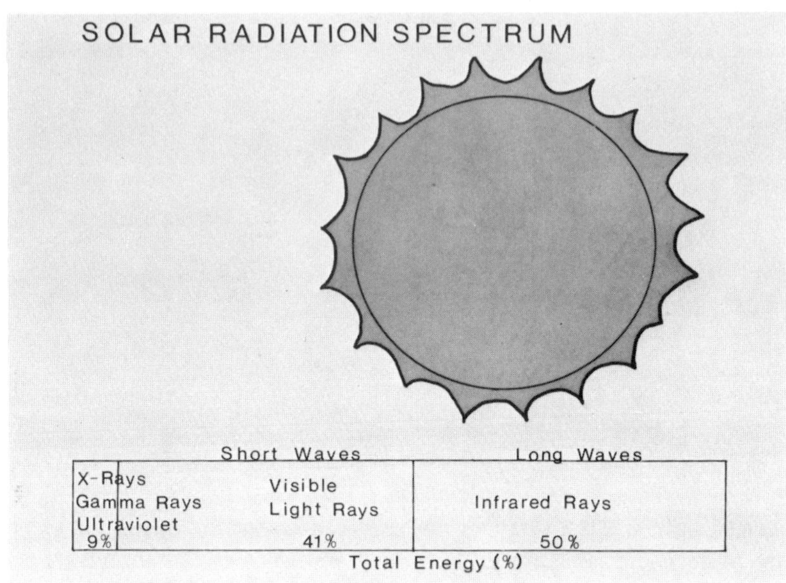

Figure 2–2

Figure 2–3

ozone. The majority of ultraviolet rays are intercepted in this fashion before they can penetrate further into the lower atmosphere. In the troposphere, water vapor and carbon dioxide are effective absorbers of the longer wavelengths (the infrared portion) of the solar radiation spectrum.

Some incoming solar radiation is *reflected* back to outer space, rather than absorbed by gas molecules. Clouds are the major reflectors in the atmosphere. They reflect back to outer space a large portion of incoming solar radiation, particularly from the visible light portion of the solar radiation spectrum. Consequently, cloudy days are much darker than clear days. On the average, clouds reflect 21% of incoming solar radiation. Thus clouds have what is called a high **albedo,** or degree of reflectivity, due partly to their water vapor composition. Water vapor is capable of absorbing part of the solar radiation spectrum, and about 3% of the total incoming solar radiation is absorbed by water vapor in clouds.

By the processes of reflection and absorption, incoming solar radiation is reduced, on the average, by 44%: 21% is reflected from clouds, 5% is reflected by dust particles, 15% is absorbed by gas molecules (and some dust particles), and 3% is absorbed by water vapor in clouds. The remaining portion of incoming solar radiation reaches the earth's surface, where even more solar radiation is lost by a process known as *surface albedo,* or the reflection of solar radiation from objects at the earth's surface. This process reduces total incoming solar radiation an additional 6%. Snow and ice are responsible for much surface albedo, for they reflect solar radiation almost as efficiently as do clouds. Thus, with 44% of the incoming solar radiation already reflected or absorbed in the atmosphere, plus another 6% reflected by surface albedo, 50% of the solar constant is lost as it passes through the atmosphere.

To summarize thus far, the portion of solar radiation that is scattered (by dust) or reflected (by albedo) returns immediately to outer space. Therefore, this portion represents an outgo of energy on the energy ledger. The portion that is absorbed by gases or clouds is temporarily stored in the earth-atmosphere system and is not lost until it is converted into heat in the atmosphere.

On the average, 50% of incoming solar radiation reaches the earth's surface, where it is absorbed. Solar radiation that has been absorbed by objects at the earth's surface is termed **insolation.** The total amount of insolation at any one location over a given period of time is dependent on the angle of the sun's rays and the length of the daylight period at that location.

Total insolation at the surface must eventually leave the earth-atmosphere system in order to maintain the earth's energy balance. Most of the total insolation is consumed and leaves the system by being converted into heat, and the remaining small portion of absorbed solar radiation is returned to outer space directly as radiant energy.

Objects that absorb solar radiation at the earth's surface experience a rise in surface temperature. Since all warm objects radiate energy, it follows that the earth's surface, in turn, radiates energy to the atmosphere. Since the temperature of objects at the earth's surface is considerably less than the temperature of the sun's surface, there is a marked contrast between the radiant energy from the sun and the radiant energy from the earth's surface. That is, solar radiation is short-wave energy, and its spectrum was previously described. Radiation from the earth's surface, by contrast, is termed **long-wave** energy, and the bulk of this radiant energy lies in the infrared portion of the energy spectrum. Of most importance in the energy balance is the ability of solar radiation to easily penetrate through the atmosphere. On the other hand, long-wave radiation does not easily penetrate the atmosphere, and thus it is temporarily retained in the earth-atmosphere system as heat. In fact, it is long-wave earth radiation that provides the majority of the heat for our atmosphere. The conversion of solar radiation into long-wave earth radiation at the earth's surface is commonly called the *greenhouse effect* because of its similarity to the heating of a greenhouse.

Now that we have seen how the conversion from incoming solar radiation to earth radiation occurs, we can examine the opposite side of the energy balance ledger, that is, outgoing energy. Long-wave radiation flows back to outer space in an amount equal to total insolation (50%): 21% returns as radiational transfers of energy, and 29% returns as nonradiational transfers of energy.

Of the 50% of solar radiation absorbed as insolation, 21% returns to the atmosphere as long-wave radiation from the surface. (See Figure 2–3.) Of this 21%, 8% escapes directly from the atmosphere without being converted into heat. The remaining 13% is actively absorbed by water vapor, carbon dioxide, methane, and nitrous oxide in the troposphere. That energy, in turn, may be reradiated back to the earth's surface by a process called **counter radiation,** and energy may be reradiated back and forth in this fashion several times before the energy is totally consumed by being converted into heat. Since counter radiation may circulate several times, it has a multiplier effect on heating the lower atmosphere.

Twenty-nine percent of insolation returns to the atmosphere as nonradiational energy transfers. Of this 29%, 9% is transferred mechanically by **conduction,** the transfer of heat by direct contact with the source of heat. Conduction is assisted by **convection**, the circulation of warm and cold air in the lower atmosphere which transfers conducted heat away from the earth's surface.

The remaining 20% of the 29% of insolation transferred in a nonradiational fashion is absorbed by the atmosphere as latent heat during the process of converting water from a liquid to a gas. Calories of latent heat are absorbed by each gram of water converted into water vapor. This heat is called **latent heat of vaporization.** It is retained by water vapor and is not released as heat in the atmosphere until the water vapor condenses into a liquid. Water vapor, therefore, is a reservoir of potential heat that is given off to the atmosphere during the condensation process.

The energy balance is now complete. Incoming solar radiation enters the earth-atmosphere system. A large portion of this energy is lost to the system by reflection and scattering. Fifty percent of solar radiation is absorbed as insolation and is eventually radiated or transferred by nonradiational methods back to the atmosphere, where it eventually escapes the system. Of all the energy received in the earth-atmosphere system, 60% is eventually converted into heat in the atmosphere.

Heat Variations on the Earth

The average temperature of the earth as a whole is relatively constant due to the balance of energy in the earth-atmosphere system. However, although the energy balance is retained over the earth as a whole, some individual latitudes receive an excess of solar radiation, while other latitudes experience a net loss of radiation during a year. In fact, poleward of about 40° latitude, more total energy is lost than is absorbed each year, whereas from 40° latitude equatorward, the opposite is true. Consequently, there is a constant transfer of energy over the earth's surface from areas of surplus to areas of deficiency. It is this transfer of energy between latitudes that is responsible for triggering the circulation of air in the atmosphere.

The variation in the composition of the earth's surface is responsible for variable heating from place to place (Figure 2–4). Land surfaces, for example, heat and cool more rapidly than water surfaces for several reasons. Land is opaque; hence, all the energy received by land is concentrated at the surface. On a clear day, therefore, land heats rapidly from the sun. At night, only a small surface area must be cooled as energy radiates back to the atmosphere. Therefore, land cools rapidly. Water, on the other hand, acts in a different fashion. Water is translucent; therefore, solar energy can penetrate water many feet in depth and is thus less concentrated than on land. However, it takes a large body of water much longer to cool down, due to the total area that must be cooled. Also, water has a much higher specific heat than land, and consequently it requires more energy to raise the temperature of one gram of water 1°F than it does to raise the temperature of one gram of dry ground. Thus, land areas generally experience greater extremes of temperature than do water areas.

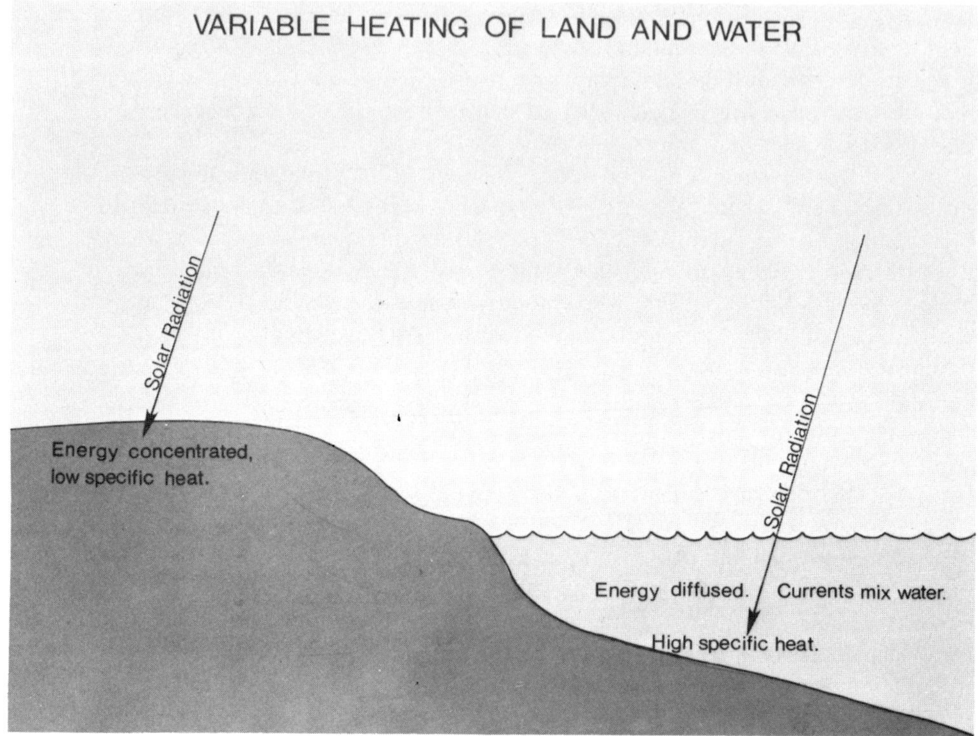

Figure 2–4

Temperature

Temperature is a measure of the amount of heat available at a given location. Temperatures vary due to earth rotation, seasonal change (which is due to revolution), latitude, and variation of altitude.

On a *seasonal* basis, temperatures vary because of the earth's revolution and the annual migration of the sun's vertical ray. On a *daily* basis (Figure 2–5), temperatures fluctuate largely with the flow of long-wave energy from the earth's surface, because long-wave energy provides the bulk of the heat to the atmosphere. The solar radiation curve and the long-wave energy curve do not coincide because part of the conversion of long-wave energy into heat is by nonradiational processes, for which a period of time elapses before the heat is released (latent heat of vaporization, for example). As a result of this time

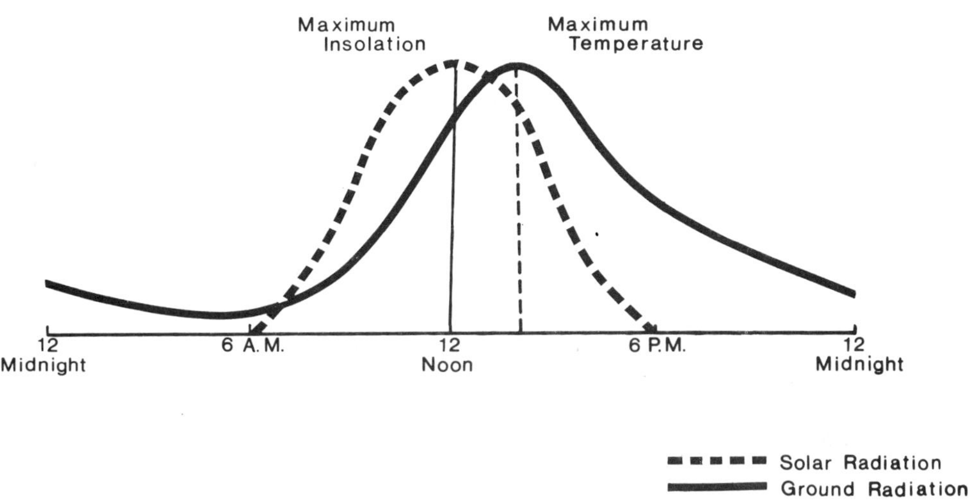

Figure 2–5

lapse, the total heat released is greater an hour or two after the peak period of absorbed solar radiation (noon). The lowest temperatures of the day are usually just prior to sunrise because by that time most of the absorbed solar radiation from the previous day has been lost from the system.

Temperatures also change with altitude. Generally, at least in the troposphere, the greater the elevation above the surface, the lower are the temperatures. Changes of temperature with altitude are called **lapse rates**. On the average, temperatures will decline at the normal lapse rate, or 3.5°F/1000 feet (6°C/1 km).

Special atmospheric conditions may cause temperatures in the troposphere to rise with increasing altitude, at least within a given zone. Increases of temperature with increases of altitude are called **temperature inversions.** Inversions may be caused by rapid temperature decline at the earth's surface due to rapid heat loss by escaping radiation, by air drainage, by colliding air masses, and by subsiding air from upper elevations. Inversions are a problem particularly when they form within several thousand feet of the earth's surface where they can prevent pollutants from rising higher into the air. When pollutants become concentrated in the lower atmosphere, they may create a potential health hazard.

Temperature Statistics

Local weather stations record temperature data several times each day. Daily temperature data are used to determine monthly and annual values from which certain implications can be drawn about seasonal variations of temperature and general climatic conditions. The **mean monthly temperature,** for example, is simply the average temperature for the month based on the average temperature for each day of the month. Likewise, the **mean annual temperature** is the average of the 12 monthly means during the year.

Temperature *ranges,* however, often provide a better indication of temperature conditions than do mean values. A temperature range measures the number of degrees that temperatures change over a period of time. The daily range, for example, is the difference between the highest and the lowest temperatures in a 24-hour period. The annual range is the difference between the average temperature of the coldest month of the year and the average temperature of the warmest month of the year.

Temperatures are shown on surface weather maps by continuous isolines called **isotherms.** Isotherms connect points of equal temperature value. (To prevent too many lines on the map, meteorologists draw isotherms usually at 4-degree intervals.) Isotherms show greater seasonal change in the Northern Hemisphere than in the Southern Hemisphere because of the greater variation of land and water surfaces in the former, which consequently causes a great variation in the concentration of absorbed solar radiation.

Heating and Cooling Degree Days

Technology has allowed us to avoid large variations in temperature by creating artificial enviroments in which to live. Homes and businesses are heated during the winter months and air-conditioned in the summer. The resulting continuous increase in demand for fuel to support heating and air conditioning is a constant source of concern to those who are in the business of supplying fuel. To assist in estimating the amount of fuel required, suppliers refer to what are termed *heating degree days* and *cooling degree days.* These indexes estimate the amount of fuel that might be consumed assuming that there will be neither a demand for heat at a daily average temperature above 65°F, nor demand for air conditioning at a daily average temperature below 70°F. If the daily average temperature is 40°F, for example, the heating degree days for that day will be 25, the difference between 40°F and 65°F. By adding daily heating degree days over an entire heating season, and averaging the total for several years, fuel suppliers can estimate approximately the amount of cubic feet of gas, or the number of barrels of oil, or the kilowatt-hours of electricity that customers whom they service will require for heating. Cooling degree days are estimated in a similar fashion during the summer. A map of the annual average heating degree days in North America is shown in Figure 2–6.

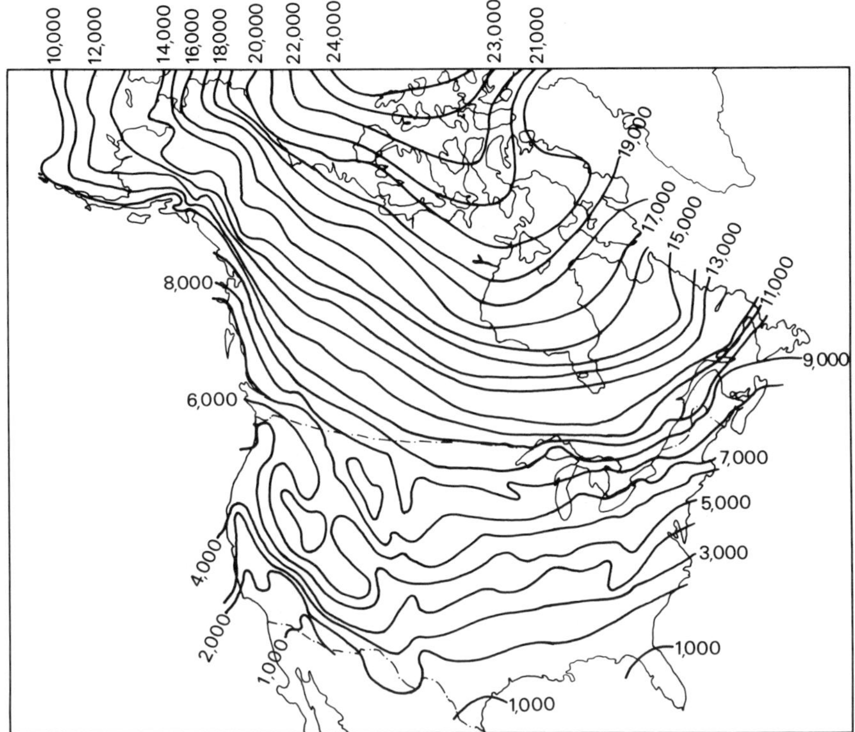

Figure 2–6

Wind Chill Factor

Another temperature-related factor that affects us is the wind chill factor. This factor is based on the sensible temperature, or the temperature felt by the exposed skin of the human body when the wind is blowing. Sensible temperatures are much lower when the wind is blowing than they are when conditions are calm. For example, a common January temperature in Minneapolis, Minnesota, is 20°F, but if the wind is blowing at a gusty 30 miles per hour, the sensible temperature plunges to −20°F. It is easy to understand the importance of covering your hands and face when temperatures are low and the wind is strong. Figure 2–7 shows a wind chill chart.

Present Temp :	+30°	+20°	+10°	0°	-10°	-20°	-30°	-40°
Wind Speed				Wind Chill				
calm	30°	20°	10°	0°	-10	-20°	-30°	-40°
5	27°	16°	6°	- 5°	-15°	-26°	36°	- 47°
10	16°	4°	- 9°	-21°	-33°	-46°	-58°	- 70°
15	9°	-5°	-18°	-36°	-45°	-58°	- 72°	- 85°
20	4°	-10°	-25°	-39°	-53°	-67°	-82°	- 96°
25	0°	-15°	-29°	-44°	-59°	-74°	-88°	- 104°
30	-2°	-18°	-33°	-48°	-63°	-79°	-94°	- 109°
35	-4°	-20°	-35°	-49°	-67°	-82°	-98°	- 113°
40	-6°	-21°	-37°	-53°	-69°	-85°	-100°	-116°

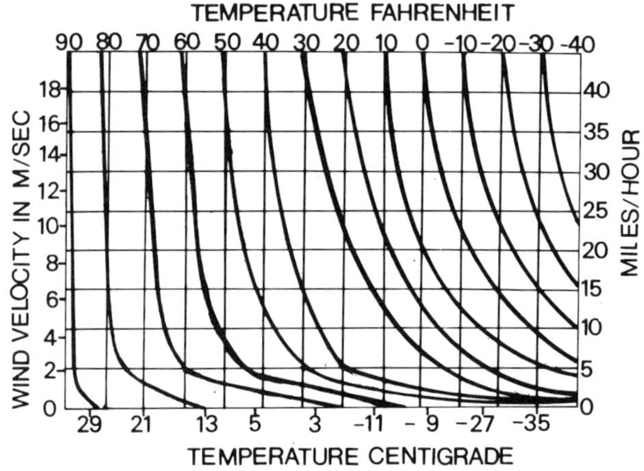

Figure 2-7

Additional Material

Temperature Scales

There are two common scales for measuring temperature. The United States generally uses the *Fahrenheit* scale, while other countries use the metric system scale, called the *Celsius* scale (formerly called *Centigrade*). On the Fahrenheit scale, the boiling point of water is 212° and the freezing point of water is 32°. On the Celsius scale, the boiling point of water is 100° and the freezing point is 0°.

One can convert Fahrenheit temperatures into Celsius equivalents by applying the following formula:

$$C = \tfrac{5}{9}(F - 32)$$

Likewise, one can convert Celsius temperatures to Fahrenheit values with the following formula:

$$F = 32 + \tfrac{9}{5}C$$

A third temperature scale is the *Kelvin,* or *absolute,* scale. It is based on absolute zero or the temperature at which a gas theoretically ceases to exert any pressure. Absolute zero is equal to $-273°C$. One degree on the Kelvin scale equals one degree on the Celsius scale. Thus, one can convert Kelvin temperature values to Celsius values merely by adding 273.

Questions

1. How does temperature vary with the layers of the atmosphere?

2. What is the composition of the atmosphere by volume?

3. What is the *spectrum* of energy waves of solar radiation? How is sunshine related to solar radiation?

4. Why does the sky generally appear blue in color?

5. What gas molecules are efficient absorbers of solar radiation?

6. Would you expect surface albedo to be higher in the Amazon rain forest or in Antarctica? Why?

7. Why do cloudy nights generally have higher temperatures than clear nights?

8. What factors affect the amount of surface insolation?

9. Why is most of the heat in the atmosphere derived from long-wave radiation and not solar radiation?

10. What is the relationship between conduction and convection?

11. What role does latent heat of vaporization play in the heating of the atmosphere?

12. How do water and land surfaces differ in heating and cooling ability?

13. Why are maximum daily temperatures usually not recorded at noon, when solar radiation is at a maximum?

14. Why do temperatures generally decrease with increased altitude in the troposphere?

15. What are the various methods by which a temperature inversion may form?

16. Is there any disadvantage to the use of mean temperature statistics?

Suggested Readings

ANTHES, R. et al. 1975. *The atmosphere*. Columbus, OH: Charles E. Merrill. Pages 20–32, 58–66.

BATTAN, L. J. 1974. *Weather*. Englewood Cliffs, NJ: Prentice-Hall. Chapter 1.

MILLER, A. 1976. *Meteorology*. 3d ed. Columbus, OH: Charles E. Merrill. Chapter 2.

MILLER, A., and THOMPSON, J. C. 1975. *Elements of meteorology*. 2d ed. Columbus, OH: Charles E. Merrill. Chapter 3.

OLIVER, J. E. 1977. *Perspectives on applied physical geography*. North Scituate, MA: Duxbury Press. Chapters 4 and 6.

STEWART, R. W. 1969. The atmosphere and the ocean. *Scientific American,* September 1969. Reprint 881. San Francisco: W. H. Freeman.

TREWARTHA, G. T. 1968. *An introduction to climate*. 4th ed. New York: McGraw-Hill. Chapters 1 and 2.

Module 3

Atmospheric Pressure and Winds

INTRODUCTION

Galileo invented the water barometer in 1593. He constructed a suction tube 34 feet long and filled it with water, but regardless of what he tried, the water would not rise above 34 feet. Torricelli in 1643 was the first to explain Galileo's problem: Water rises in a suction tube placed in a tank of water due to the weight of the air pressing down on the water in the tank. Until this finding, people believed that air was a weightless substance. Torricelli, who was a student of Galileo, also hypothesized that a liquid denser than water would not require a tube as long as the water barometer to show changes in air pressure. In further experimentation, he selected mercury as a denser liquid. The mercury barometer was born, and it required a tube only slightly longer than 30 inches.

Subsequent experiments with Torricelli's mercury barometer showed that air pressure decreases with increases in elevation, but it was not until the seventeenth and eighteenth centuries that scientists realized that changes in air pressure were related also to weather. In 1661, Robert Boyle discovered the relationship between pressure and density. In 1802, Jacques Charles discovered the relationship between temperature and density in a gas. Shortly thereafter, Joseph Gay-Lussac demonstrated the relationship between pressure with changes in the volume of a gas. The relationship between pressure and winds was explained later.

Until the 1800s, people thought that storms both began and died in the same area. Benjamin Franklin was one of the first to suggest that air travels great distances and that storms in one area may travel to another area with the winds. In 1883, the eruption of the volcano Krakatoa, on an island near Java, threw volcanic dust into the air, traces of which were found 8000 miles away, in England, only a few days later. The dust was responsible for creating some brilliant sunsets for several months. Early sailors learned quickly that at certain latitudes winds blow from generally the same direction. In fact, the names for the wind belts today are derived largely from the names given to them by the early sailors: the *trade winds,* which blow from east to west, aided trading vessels going to the New World; the *Roaring Forties,* which blow consistently from west to east and include many storms; and the *Screaming Fifties,* also known for storminess and strong winds, which blow from west to east. From this type of evidence it became increasingly apparent that air in the atmosphere circulates great distances around the earth. In fact, meteorologists still do not completely understand the relationship between surface air circulation and upper-air jet streams, which were not discovered until World War II.

In light of the energy crisis of the 1970s, the idea of harnessing the wind to generate electricity is becoming more popular. In the first one-third of this century, the Great Plains were dotted by windmills which generated the current to operate the pumps that lifted water from wells. With the advent of cheap electricity in the late 1930s, most of these original windmills were abandoned. Only in the last few years have windmills increased in popularity. In fact, experimental windmills are now being constructed in the hope that they may become efficient enough to supply a significant portion of the electricity needed in homes.

OBJECTIVES

By the end of this module, you should be able to do the following:

1. Explain how air pressure is measured and how it varies with altitude.

2. Describe the factors that cause changes in air pressure.
3. Explain the cause and the characteristics of the world pressure belts.
4. Explain the factors that cause winds to vary in velocity and direction.
5. Explain and describe the circulation of air in low- and high-pressure cells.
6. Explain the causes and the characteristics of the world wind belts.
7. Explain how the differential heating of land and water affects air circulation.
8. Explain and describe specialized local winds.

KEY TERMS

atmospheric pressure

barometer

millibar

isobar

pressure ridge

pressure trough

advection

pressure gradient force

Coriolis force

geostrophic balance

geostrophic wind

friction

cyclones

anticyclones

polar front

monsoon

katabatic wind

foehn wind

Chinook wind

Now you are ready to begin the audiovisual portion of this module. Select the MEDIAPAK 3 component(s) and proceed. Following is a topical outline of the audiovisual sequence. You will find this outline helpful for reference and review. After completing MEDIAPAK 3, return to this book to perform the exercises.

OUTLINE

Atmospheric pressure
Measurement
Factors for pressure change
World pressure belts

Wind
Factors that control wind velocity and direction
Circulation of air in low- and high-pressure cells
World wind belts
Important local winds

Exercises

1. Which of the following is not a control of air pressure?
   A. Temperature
   B. Altitude
   C. Changes in volume
   D. Friction

2. 1015.5 millibars is equivalent to _________ inches of mercury.
   A. 30.00
   B. 29.92
   C. 29.90
   D. 31.05
   E. 30.50

3. On a day-to-day basis, low-pressure cells may be located within the subtropical high-pressure belt.
   A. True
   B. False

4. The subpolar low-pressure belt is characterized by
   A. year-round low-pressure cells over landmasses in the Northern Hemisphere.
   B. The Aleutian low over the Pacific Ocean and the Icelandic low over the Atlantic Ocean.
   C. consistently well-developed low-pressure cells over the entire Southern Hemisphere at these latitudes.
   D. a latitude range of 25° to 85°.

5. The following diagram is typical of the circulation in a ___________-pressure cell.
   A. high
   B. low

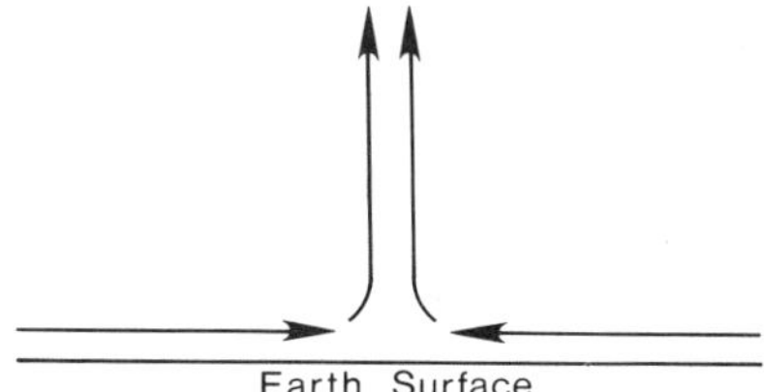

6. The closer the spacing of isobars on a surface weather map, the ___________ is the pressure gradient force.
   A. greater
   B. less

7. The Coriolis force causes objects to be deflected to the ___________ in the Northern Hemisphere.
   A. right
   B. left

8. A geostrophic balance is reached when the
   A. pressure gradient force equals the friction force.
   B. pressure gradient force equals the Coriolis force.
   C. Coriolis force equals the friction force.
   D. pressure gradient force, the Coriolis force, and the friction force are all equal.

9. The trade winds blow from the _________________ to the _________________.
   A. subpolar low; equatorial high
   B. polar high; subpolar low
   C. subtropical high; equatorial low
   D. subtropical high; subpolar low
   E. doldrums; horse latitudes

10. A land breeze is the result of high pressure on the land and lower pressure over adjacent water.
   A. True
   B. False

11. The chinook is an example of a
   A. monsoon wind.
   B. foehn wind.
   C. katabatic wind.
   D. sea breeze.
   E. valley breeze.

Summary

Atmospheric Pressure

The atmosphere exerts a force on all objects on the earth's surface due to gravity. At sea level the atmosphere is thick enough to exert a pressure on all objects of 14.7 lb/in^2. This so-called **atmospheric pressure** is equal to the amount of weight exerted on a part of the earth's surface by the weight of a column of air that extends to the top of the atmosphere.

Atmospheric pressure decreases as height increases. In fact, above 18,500 feet (5500 m), pressures drop very rapidly, and above 100,000 feet they are almost too small to measure. Atmospheric pressure can be measured with a **barometer.** Barometers are basically of one of two types: mercury and aneroid. Barometers accurately measure changes in pressure in inches of mercury, or **millibars.** One inch of mercury is equivalent to 33.85 millibars.

Variations in pressure from place to place are easily noted by analyzing the pattern of pressure values recorded on a surface weather map. On such maps, **isobars** are used to connect points of equal pressure. Typically, isobars appear in circular patches of high or low pressure called *pressure cells*. Pressure cells vary in size and shape. On the average, individual cells have diameters of 500 to 1500 miles. High- and low-pressure cells that have elongated shapes are called **pressure ridges** or **pressure troughs,** respectively.

Air pressure varies from place to place and with time, for as a gas, air conforms to the natural physical laws that govern all gases. Pressures may increase or decrease with changes in temperature or volume. Boyle's law states that if the temperature of a gas does not change, its density (or number of molecules per unit volume) varies directly as the pressure varies. In other words, pressures may increase or decrease if there is a density change, provided temperature remains constant. Charles's law states that if the pressure within a gas is unchanged, its volume will change in proportion to any temperature change that may occur. In other words, pressures may increase and decrease with changes in temperature, provided there is a change in volume. Combining the gas laws implies that pressure, temperature, and volume are completely interdependent, such that a change in one of them will cause a compensating change to occur in one, or both, of the remainder:

Boyle's law:

At a constant temperature, the volume (V) of a mass varies inversely as its pressure (P). That is,

$$P = \frac{K_1}{V} \qquad \text{where } K_1 = \text{a constant}$$

Charles's law:

At a constant pressure, volume varies directly with temperature (T). That is,

$$V = \frac{K_2}{T} \qquad \text{where } K_2 = \text{a constant}$$

In summary, atmospheric pressures vary because of changes in *temperature, volume,* or *altitude.* Thus, it is possible for two places to have the same temperature but different pressures due to adjustments in the volume of air. Likewise, two areas may have identical pressures but different temperatures due to changes in volume. Finally, temperatures may change and cause changes in pressure because the volume of the air remains constant but the density of the air changes.

World Pressure Belts

When average atmospheric pressure values are analyzed on a worldwide basis, they reveal a definite pattern of pressure such that within certain latitudinal boundaries, the average pressure is high or low. Enlarged areas of the same average pressure are called *pressure belts*. The world pressure belts represent *average* conditions, not actual daily conditions. As the foundation for understanding the world pattern of moisture and air circulation, however, they are useful knowledge.

Several basic statements can be made about the world pressure belts:

1. They all represent mean sea-level pressure.
2. They are not representative of day-to-day conditions.
3. They all shift poleward and equatorward in flux with the shifting vertical ray of the sun.
4. Boundaries between belts are broad transition zones, not sharply defined lines.

There are four major pressure belts (see Figure 3–1), and they are located in both the Northern and the Southern hemispheres: the equatorial low, the subtropical high, the subpolar low, and the polar high. The *equatorial low belt* appears to be associated with the high temperatures characteristic of latitudes equatorward of 15°. Warm air is less dense than cold air and therefore exerts less pressure. The *subtropical high belt* extends from about 15° to 35° latitude. The presence of high pressure in a region of relatively high year-round temperatures leads to the conclusion that this belt is closely associated with the sinking of air from aloft in this region, which causes the lower surface air to be compressed, resulting in turn in high pressure. The *subpolar low belt*

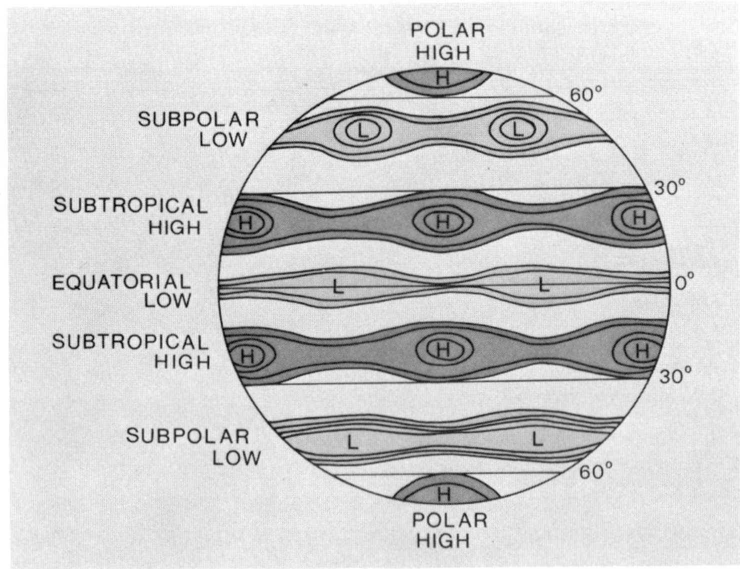

Figure 3–1

extends from about 35° to 70° latitude. This belt appears to be associated with temperature changes. In the summer months high temperatures favor low-pressure development, but in the winter the belt is considerably weaker as cold, dense air increases pressures. In the Northern Hemisphere low-pressure cells in the subpolar low are found year-round only over the oceans. In the Southern Hemisphere low-pressure cells are weak all year due to the absence of major landmasses. The *polar high belts* extend poleward from 70° latitude. Like the subpolar lows, the polar highs vary with the seasons, having their best development in spring and summer.

The significance of the world pressure belts is the alternating pattern of highs and lows, because winds in the atmosphere are triggered by pressure differences, as we will demonstrate in the next section.

Wind

The transfer of energy between latitudes, the related variations in heating at different latitudes, and the differences in pressure from place to place—all are responsible for triggering the movement of air, or wind, in the atmosphere. Wind, or **advection,** is the horizontal motion of air over the earth's surface. Winds blow with changing velocity and direction.

In order for an object or a mass to be moved, some force must push or pull it, and wind is no exception. That is, air must be activated by a force to become wind. Newton's second law summarizes the relation between the force necessary to move an object at a particular rate of speed: The mass of an object times a force is proportional to the acceleration (change in velocity) of the object. In the atmosphere, the force that accelerates air is atmospheric pressure.

If two unequal and opposing forces are applied to an object, the object will be moved in the direction of the lesser force. In the atmosphere, differences in atmospheric pressure apply forces to air, and, as a result, air always moves from a region of higher pressure to a region of lower pressure. This force, which is the result of pressure differences, is called the **pressure gradient force.** The greater the difference in pressure over a given distance, the greater is the pressure gradient force. In Figure 3–2, the pressure gradient force is twice as great on the air parcel on the left as in the diagram on the right, because the pressure difference on each side of the air parcel is twice as large. As a result, the air parcel accelerates at twice the velocity in the left diagram. The closer the spacing of isobars (equal-pressure lines), the greater will be the acceleration or velocity of the wind.

Pressure gradient force causes air to move, but air does not flow directly from high to low pressure near the earth's surface because the earth's surface is constantly rotating beneath the atmosphere. Because of earth rotation, air

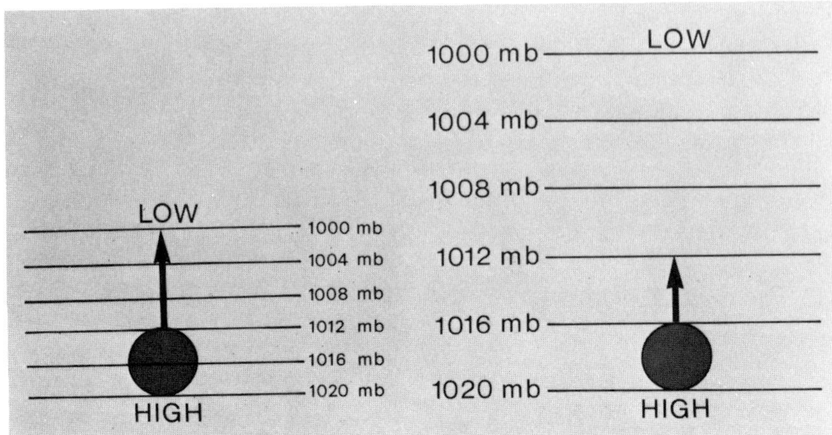

Figure 3–2

regions. This deflection of wind is termed the **Coriolis force.** The earth
rotates from west to east, and therefore winds are deflected to the right in the
Northern Hemisphere. Thus, so far we have seen how two forces act on an air
parcel: the pressure gradient force *and* the Coriolis force. Overall, the result
(in the Northern Hemisphere) is that the Coriolis force acts at right angles to
the direction of motion and the air assumes a direction of motion at some
angle across the isobars (see Figure 3–3).

The Coriolis force increases with increasing air velocity due to angular
momentum of the air, and it increases with increasing latitude. Assuming that

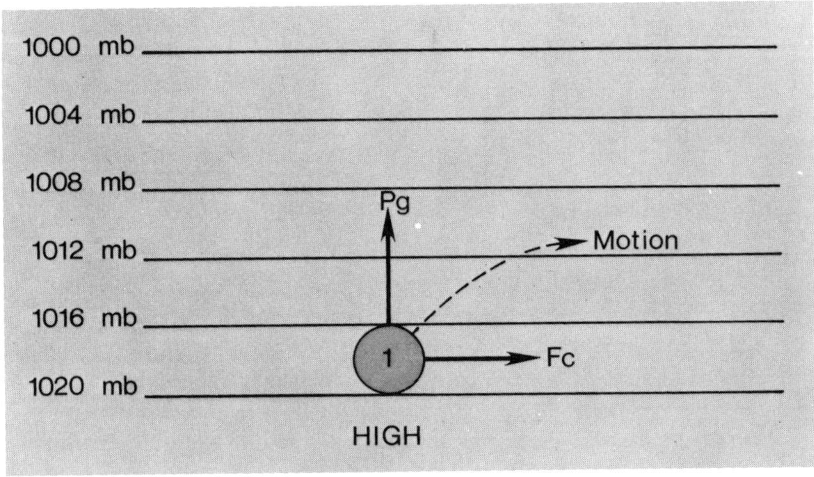

Figure 3–3

no other factor were involved, the Coriolis force and the pressure gradient force would eventually become equal and directly opposed to one another (see Figure 3–4). Consequently, there would be no net force on the air parcel and it would travel indefinitely in a straight path between isobars. This balance is called a **geostrophic balance,** and the air motion is called a **geostrophic wind.**

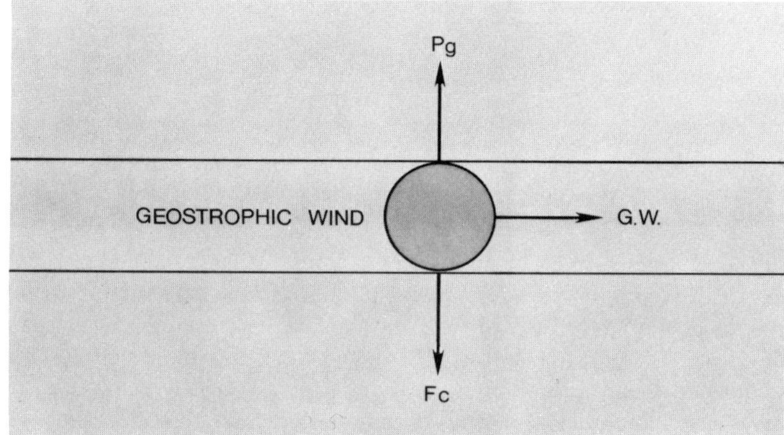

Figure 3–4

Based on the pressure gradient force and the Coriolis force, the direction of flow of an air parcel in the Northern Hemisphere is counterclockwise around a low and clockwise around a high. In the Southern Hemisphere the circulations are reversed. Thus, if you stand outside with your back to the wind, your left side will always point toward the direction of lower pressure (Northern Hemisphere).

A third and essential factor that affects winds is **friction,** the degree of resistance that an object meets as it moves over another object. Without friction, the circulation of air would always be geostrophic, and active weather could not form, for air must flow out from the center of high-pressure cells into the center of low-pressure cells for active weather to exist. The effect of friction on wind is to reduce its velocity. Once the velocity is reduced, there is a corresponding reduction in the Coriolis force. As a result of reduced Coriolis force, the pressure gradient force exceeds the Coriolis force, and the air parcel moves across the isobars (see Figure 3–5). Once again, the direction of air circulation is from regions of high pressure to regions of low pressure. Friction is most effective near the earth's surface, particularly over land surfaces and below 3300 feet (1000 m) in elevation. Upper-air winds, in contrast, tend to flow geostrophically most of the time.

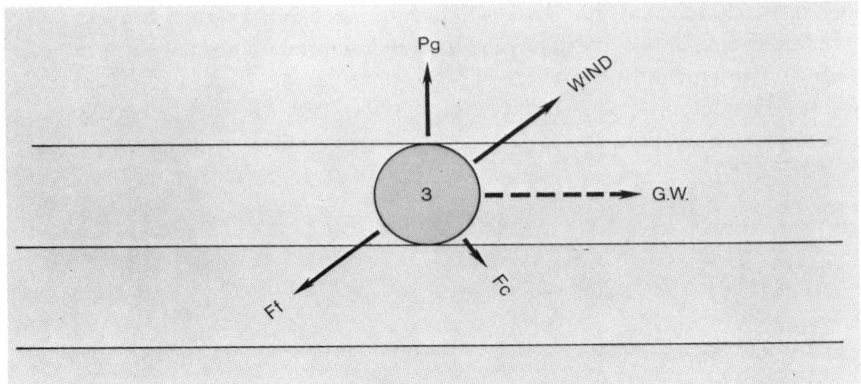

Figure 3–5

The circulation of air in surface low-pressure cells in the Northern Hemisphere (due to pressure gradient force, Coriolis force, and friction force) is counterclockwise and converges into the center of the cell. Once air converges at the center of a low, it has no place to go but up; thus, rising air is also characteristic of air circulation in a low-pressure cell. This type of circulation is called *cyclonic* circulation, and for that reason, low-pressure cells are also called **cyclones** (see Figure 3–6A).

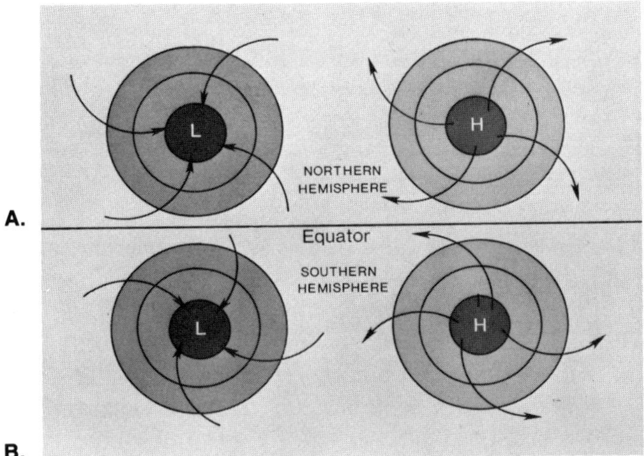

Figure 3–6

The circulation of air in a high-pressure cell is clockwise, diverging, and sinking in the center. This circulation is called *anticyclonic,* and hence high-pressure cells are also called **anticyclones** (see Figure 3–6B).

Winds are named after the direction from which they originate. A south wind, for example, blows from the south, and a westerly wind blows in a generally west-to-east direction.

World Wind Systems

There is a worldwide system of winds that are initiated by the variations in pressure between the world pressure belts (see Figure 3–7). These planetary winds, by and large, reflect the broad-scale circulation of air in the troposphere.

WORLD WIND SYSTEMS

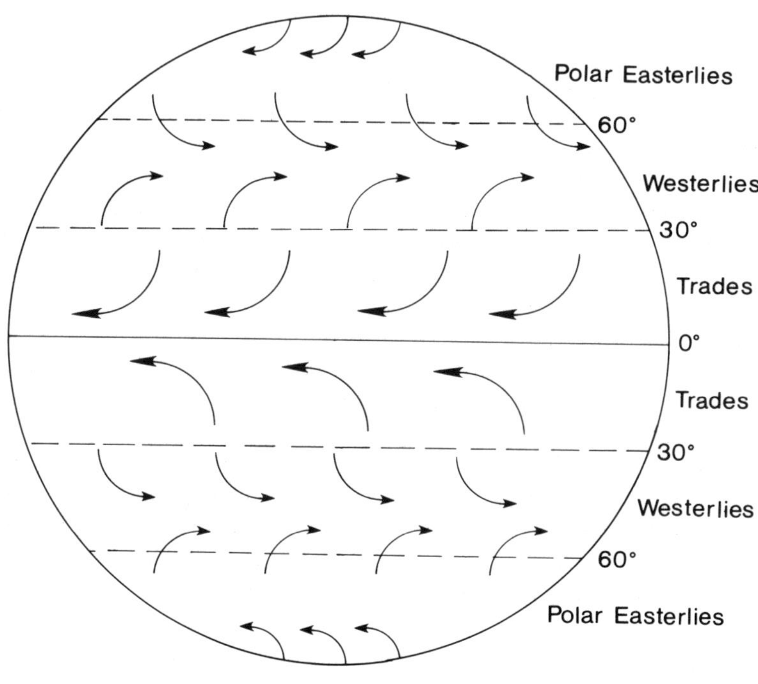

Figure 3–7

All winds blow from high- to low-pressure regions. Therefore, you must look to the subtropical highs and the polar highs as the starting point for the planetary winds. One set of winds blows from the subtropical highs into the equatorial lows. These winds are easterly winds and are called the *tropical easterlies* or trade winds. A second set of winds blows from the subtropical highs into the subpolar lows. These winds are westerly and are, in fact, referred to as the *westerlies*. Another set of planetary winds emanates from the polar highs and flows into the subpolar lows. These relatively weak easterly winds are termed the *polar easterlies*.

Daily weather systems flow around the world directed by the circulation of air within these general world wind systems. The United States, for example, lies astride the middle latitudes in the heart of the westerly wind belt. Weather systems, therefore, circulate across the country in various patterns from the west coast to the east coast. Indeed, most weather observers look generally to the west to view changes in weather. If you lived within 15° of the equator, however, new weather systems would be forthcoming generally from the east.

Local Winds

The world wind systems dictate the general flow of surface air. Local variations of wind, however, may have a much greater impact on daily weather in an area.

Areas where land and water adjoin often experience day-to-day winds called *land* and *sea breezes* (see Figure 3–8). Land and sea breezes are caused by the pressure gradient that develops between land and water due to temperature differences. A warm land surface adjacent to a cooler water surface on a clear day may result in greater pressure over the water than the land due to the relationship between air temperature and density. As a result, air will flow from the sea onshore, creating a cooling sea breeze. At night the land may be cooler than the water. Therefore, the pressure gradient will be reversed, and winds will then flow from land to sea and form a land breeze.

Pressure variations between adjoining land and water areas can operate on a larger scale and over longer time periods. Such is the case in the formation of **monsoon winds.** Warm land surfaces in summer will assist the development of low-pressure cells over the land, while, simultaneously, slightly cooler water surfaces will lead to denser air and higher pressures over the water. During the summer, monsoon winds blow onshore and often bring with them large amounts of evaporated moisture which is often condensed into rain over the land. In the winter the pressure situation is reversed, with high pressure that dominates over the cold land surfaces. The wind direction, therefore, is reversed; that is, the winter monsoon blows offshore and often is associated with a dry season. Monsoon winds are classically associated with Southern Asia, but almost every continent experiences some degree of monsoonal air circulation.

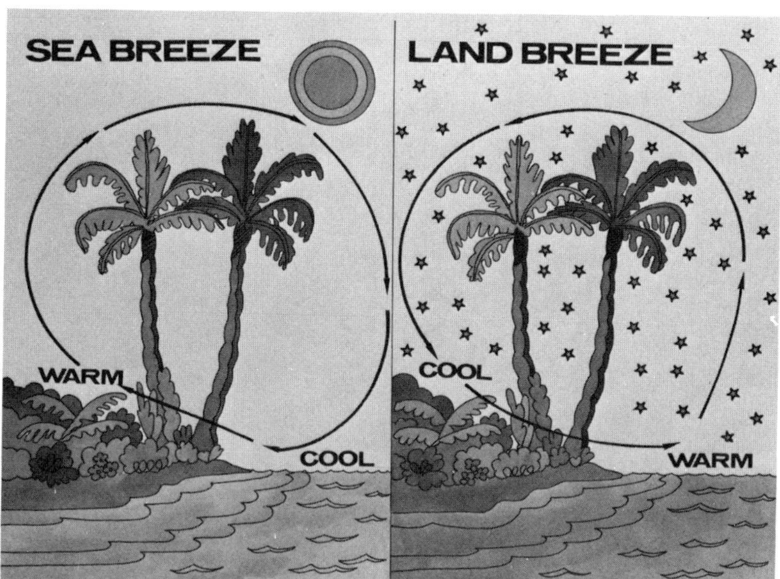

Figure 3–8

Regions of mountainous terrain experience local winds called *mountain* and *valley breezes*. Once again, these local winds are triggered by pressure differences created by temperature variations. During daytime hours, heated mountain slopes cause winds to flow upslope from cooler valleys; these winds are called *valley winds*. In the evening, the upper mountain slopes cool rapidly, and the air is denser than in the valleys. Consequently, the wind direction is reversed, and wind flows into the warmer valleys; this breeze is the *mountain breeze*.

Katabatic winds are the result of cold, dense air flowing downslope by gravity. Most common in winter, these winds are usually very gentle, but they may be of very high velocity when they are confined to narrow channels, as in a canyon.

Foehn winds are the result of prevailing winds that are forced to rise over mountains in their paths. As the winds flow down the far side of a mountain, they are compressed and heated. This warm, dry wind on the lee side of mountains is the foehn wind. On the eastern side of the Rocky Mountains, these winds are referred to as **chinook winds,** after the Chinook Indians.

The circulation of air in the atmosphere helps to maintain the balance of energy in the earth-atmosphere system. Excess solar energy absorbed at the equatorial latitudes is transferred to more poleward latitudes, where a net loss of energy is an annual characteristic. Air rises into the upper troposphere over the equatorial regions, where it is circulated poleward by upper-air winds to approximately 30° latitude, where the air cools sufficiently to sink back to the surface. Thus, a giant convection system rotates air in the troposphere between the equator and 30° lattitude. Poleward of 30°, upper-air winds transport some energy from the equatorial regions into higher latitudes. Sinking air in the polar regions tends to flow equatorward into the middle latitudes, where it is warmed. A second convection system is thus established between 30° and 90° latitude.

Questions

1. What is the relationship between pressure and density?

2. What is the value of placing isobars on surface weather maps?

3. Under what conditions can air be compressed in the atmosphere?

4. Why does the relationship between temperature and pressure not remain constant?

5. What general statements can be made about the world pressure belts?

6. How does air circulation in the upper atmosphere relate to the subtropical high-pressure belt?

7. What is the relationship between pressure gradient force and wind velocity?

8. How does friction assist in causing air to circulate into low-pressure cells and out of high-pressure cells?

9. How do you explain the seasonal reversal of monsoon winds?

10. What is the relationship between the world pressure belts and the world wind belts?

11. Would you expect mountain breezes during the day or at night? Explain.

Suggested Readings

BATTAN, L. J. 1974. *Weather*. Englewood Cliffs, NJ: Prentice-Hall. Pages 30–52.

LEHR, P. et al. 1975. *Weather*. New York: Golden Press. Pages 53–59.

McDONALD, J. E. 1952. The Coriolis effect. *Scientific American,* May 1952. Reprint 839. San Francisco: W. H. Freeman.

MILLER, A. 1976. *Meteorology*. 3d ed. Columbus, OH: Charles E. Merrill. Pages 51–62, 75–86.

MILLER, A., and THOMPSON, J. C. 1975. *Elements of meteorology.* 2d ed. Columbus, OH: Charles E. Merrill. Pages 83–101, 115–39.

MILLER, P. J.; BRINKMANN, W. A. R.; and BARRY, R. G. 1974. Windstorms: a case study of wind hazard for Boulder, Colorado. In *Natural Hazards,* by G. White. New York: Oxford University Press. Pages 80–86.

TREWARTHA, G. T. 1968. *Introduction to climate.* 4th ed. New York: McGraw-Hill. Chapter 3.

Module 4

Atmospheric Moisture and Precipitation

INTRODUCTION

World population continues to increase at an alarming rate, and as the number of people increases, so also does the demand for fresh water. In many areas of the world the demand for fresh water greatly exceeds the supply. Periodic droughts which may have created a temporary inconvenience 30 or 40 years ago often become disasters today because of the great increase in demand for water. For example, in the early 1970s, the people and animals living in the lands bordering the southern edge of the Sahara Desert fought starvation and death due to a drought that lasted several years. Also, much of California was forced to impose severe water rationing in 1977 due to the lack of normal winter rains. Increasingly we hear of wells drying up, streams and rivers running dry, and groundwater supplies rapidly diminishing. Yet despite these problems, people continue to expand into regions of marginal land where water supplies are low. Very few people, for example, voice concern over the expansion of retirement communities in the deserts of the Southwest. The shortage of water in many Arab lands is so acute it has prompted some Arab leaders to seriously consider cutting away huge icebergs from Antarctica and transporting them by sea to Arab lands where they can be melted into fresh water. The problem of adequate fresh water may possibly become the water crisis of the 1980s, much as adequate energy is the crisis of the 1970s.

At times, water may be more abundant than is desirable, and floods, which occur periodically in most parts of the world, may result. Flood-control programs that feature special dams and containment structures are common along many rivers and streams. In Holland, the battle between sea and land has been waged for many centuries, and land there has been gradually captured from the sea through the construction of large and extensive dikes. However, our ability to control the flow of water over the earth's surface is still far from complete. If we could entrap excess water from one place and conveniently move it to those regions where supply is short, impending crises could largely be avoided. Obviously, we are a long way from controlling the world's water. In order to control water, we would have to control the surface distribution of water and be able to constantly monitor the formation of moisture in the atmosphere.

OBJECTIVES

By the end of this module, you should be able to do the following:

1. Explain the process of evaporation and the factors that assist the process.
2. Briefly describe the flow of moisture within the hydrologic cycle.
3. Explain the relationship between temperature, water vapor capacity, saturation, and vapor pressure.
4. Explain specific humidity, absolute humidity, and relative humidity.
5. Calculate relative humidity when given a list of temperatures and water vapor capacities.

6. Explain the condensation process.
7. Explain the relationship between the condensation process and the dew point.
8. Explain adiabatic temperature change.
9. Explain the mechanisms that produce raindrops.
10. Describe the three methods by which large masses of air are forced to rise and cool.
11. Briefly describe the world distribution of precipitation.
11. Discuss the shortcomings of average annual precipitation values and the importance of seasonal distribution.

KEY TERMS

humidity	specific humidity
water vapor	relative humidity
evaporation	dew point
transpiration	condensation nuclei
potential evaporation	sublimation
latent heat of vaporization	adiabatic temperature change
vapor pressure	latent heat of condensation
water vapor capacity	condensation level
saturation	windward slope
absolute humidity	leeward slope

Now you are ready to begin the audiovisual portion of this module. Select the **MEDIAPAK** *4 component(s) and proceed. Following is a topical outline of the audiovisual sequence. You will find this outline helpful for reference and review. After completing* **MEDIAPAK** *4, return to this book to perform the exercises.*

OUTLINE

Humidity
Evaporation process
Hydrologic cycle
Measures of water vapor
Condensation process
    Adiabatic temperature change

Precipitation
Mechanics of raindrop formation
Snow, sleet, and hail
Convectional, orographic, and cyclonic precipitation
World distribution pattern
Average and seasonal precipitation

Exercises

1. Evaporation amounts will be high when
    A.   the temperature is high.
    B.   winds have high velocity.
    C.   the initial amount of water vapor in the air is low.
    D.   the supply of moisture is high.
    E.   all of the above apply.

2. In the hydrologic cycle, more moisture is transferred as water vapor in the atmosphere than by direct runoff from the land.
    A.   True
    B.   False

3. The higher the air temperature, the _______________ is the air's capacity to hold water vapor.
    A.   greater
    B.   less

4. If a cubic meter of air at 50°F (10°C) has a capacity of 9.6 grams of water vapor, and it is presently holding 2.4 grams, the air's relative humidity is _______ percent.
    A.   50
    B.   25
    C.   40
    D.   10
    E.   75

5. When air is cooled to the dew point,
    A.   the relative humidity reaches 100%.
    B.   latent heat of condensation is released to the air.
    C.   it has reached the condensation level.
    D.   vapor pressure reaches a maximum and the air is saturated.
    E.   all of the above apply.

6. The dry adiabatic lapse rate applies whenever
    A.   the air is saturated.
    B.   the air rises above the condensation level.
    C.   the air temperature is higher than dew point.
    D.   the relative humidity is 100%.
    E.   all of the above apply.

7. The Bergeron process is based on the availability of different-sized water droplets in a cloud.
    A.   True
    B.   False

8. With orographic precipitation, the majority of total rainfall is deposited on the
   _________________ slope.
   A. windward
   B. leeward

9. Which of the following statements is true?
   A. Precipitation amounts generally increase from the equator poleward.
   B. Coastal areas near cold ocean currents usually are drier than coastal areas
      adjacent to warm ocean currents.
   C. Precipitation generally increases the further you move into the interior of a
      continent.
   D. Precipitation is usually higher on leeward slopes than on windward slopes.
   E. None of the above.

10. Deviations from seasonal precipitation norms are more critical to agriculture in
    regions where annual precipitation totals are high than in regions where annual
    precipitation amounts are low.
    A. True
    B. False

Summary

Water is a unique substance in that it can exist in three different states under
normal atmospheric conditions: It can be a gas, a liquid, or a solid. It freezes
at 0°C (32°F), and at almost all temperatures it can be evaporated as a gas. For
almost any other substance to change state, there must be much greater variations of heat or cold than those occurring naturally in the atmosphere.

Humidity is a measure of the amount of moisture in the atmosphere. The
term generally refers to the amount of water vapor (a gas) or liquid droplets on
solid objects, and it is generally separated from the various forms of precipitation. **Water vapor,** or water in gaseous form, supplies all the moisture for
precipitation. Therefore, it is important to know both how water vapor gets
into the air and also how water vapor combines to form precipitation.

The process of converting water into water vapor is called **evaporation.**
In order for evaporation to occur, there must be an exchange of energy. The
energy for this process is heat energy from the sun. In the process of evaporation, calories of heat energy are removed from the evaporating liquid surface.
As a result, the evaporating surface is cooled. The process is comparable to
the perspiring of your body. As you perspire, calories of heat are given off and
your body is cooled. The calories of heat consumed in the evaporation process
are absorbed by water vapor in the air in the form of latent heat, called **latent**

heat of vaporization. Latent heat of vaporization absorbed by the air does not result in a temperature change in the air. Instead, this latent heat absorbed by water vapor represents a reservoir of **potential** heat that is realeased during the conversion of water vapor back into a liquid or a solid.

Evaporation is most active under the following conditions:

1. when temperatures are high;
2. when the air has little water vapor initially;
3. when there are strong winds to mix the air; and
4. when there is an ample supply of water to be evaporated.

The ability of air to evaporate water is called **potential evaporation.** Often potential evaporation exceeds the amount of moisture actually evaporated. This situation is particularly common in warm desert regions where the available supply of moisture is generally very low.

Total evaporation includes a considerable supply of water vapor derived from the surfaces of plant leaves; this form of evaporation is called **transpiration.** A large tree on a hot summer day, for example, may transpire several hundred gallons of water into the air.

Condensation, the process of converting water vapor into a liquid, also requires an energy exchange. In the condensation process the calories of latent heat of vaporization held in the water vapor are released to the surrounding air. This released heat is called the **latent heat of condensation.** Unlike the latent heat of vaporization, the latent heat of condensation does result in an increase in air temperature. The condensation process is thus a warming process.

Water is continually being converted from a liquid or solid to a gas and back again. The existence of water in its various states in the earth-atmosphere system follows a cyclical pattern called the *hydrologic cycle* or *water cycle.* The water cycle involves the basic exchange of moisture between the earth's surface and the atmosphere by evaporation, condensation, and precipitation. The two major subdivisions in the cycle are the transfer of moisture (1) between land and sea and (2) between latitudes.

The moisture exchange between land and sea is essential in order to maintain a moisture balance because on land surfaces, precipitation exceeds evaporation, and over the sea, evaporation exceeds precipitation. Moisture evaporated from the sea is transported over the land, where it condenses and falls as precipitation. Of the moisture that reaches the earth's surface, some is absorbed into the ground as a groundwater resource; another portion runs off the surface as streams and rivers and is returned to the sea. The majority of moisture on the earth's surface, however, returns to the atmosphere by evaporation from the land, some of which is transported by winds, where it condenses into precipitation and falls on other land areas or precipitates into the sea.

The significant factor in the water cycle is that tremendous volumes of moisture are transported back and forth in the atmosphere in the form of water vapor. This moisture is transported not only from land to sea but also across latitudes.

Measures of Humidity

The amount of water vapor that the air can hold is determined by its temperature. The higher the temperature, the greater is the air's ability to hold water vapor. At any given temperature, the amount of water vapor that the air can hold is fixed and is termed the **water vapor capacity.** Water vapor capacity increases with temperature at an increasing rate.

Several methods are used to express the amount of water vapor in the air. **Absolute humidity** refers to the actual weight of water vapor contained in a given volume of air. Absolute humidity, however, is not a commonly expressed measure of water vapor because as air moves vertically in the atmosphere, its volume constantly changes, causing a constant change in the absolute humidity value. To avoid that problem, meteorologists use the measure of **specific humidity,** the weight of water vapor *per unit* weight of air. It is expressed most commonly as grams per kilogram of moist air. Should the volume of air change but the amount of water vapor remain constant, the specific humidity will also remain constant.

A third expression of water vapor is the **relative humidity.** The relative humidity is the ratio between the amount of water vapor actually in the air at a given temperature compared to the maximum amount that the air can hold at that temperature. Relative humidity is expressed as a percentage. For example, if a cubic meter (1 m^3) of air holds 4 grams of water vapor at a given temperature, and the capacity of the air at that temperature is 8 grams, the air is holding only one-half the amount of water vapor that it could; the relative humidity of that air is, therefore, 50%.

Condensation and Precipitation

How does water vapor condense into clouds and produce precipitation? Air acts much like a sponge. It continues to absorb water vapor at a given temperature until it reaches capacity or is **saturated.** Additional water vapor added to saturated air is an excess, and it is this excess that condenses into minute liquid water molecules. Condensation and precipitation will develop only when the air is saturated.

This process appears to be a simple one; that is, air simply absorbs amounts of water vapor over an area until it is saturated. Then the water vapor condenses and precipitates back to earth. Rarely, however, does the air absorb sufficient amounts of water vapor for precipitation to occur, because air does not remain stationary over a source of moisture long enough. Consequently, air must become saturated in another manner.

Two factors control the degree of saturation of air: (1) the amount of water vapor in the air and (2) the air temperature. Figure 4–1 is a chart of water vapor capacities at various temperatures. At 10°C, air has a capacity of 9.6 grams/cubic meter (9.6 g/m^3). Let us assume that a parcel of air contains 6.8 g/m^3 at 10°C. This parcel of air is not saturated, but it can become saturated in one of two ways. One method is to add an additional 2.8 grams of water vapor to the 6.8 grams already in the air; the total amount would equal 9.6 grams, which is the air's capacity at 10°C. But what if there is no additional water vapor available? A second method of reaching saturation can be used; that is, the temperature of the air can be lowered from 10°C to 5°C. At 5°C, the capacity of the air is 6.8 g/m^3. Since there are already 6.8 grams/m^3 of water vapor in the air, the air is now at capacity or saturated. Consequently, it is possible to saturate air by cooling it.

RELATIVE HUMIDITY

WATER VAPOR
  CAPACITY

Temp. (°C)	Grams/ m^3
−15	2.0
−10	2.2
−5	2.8
0	4.2
+5	6.8
+10	9.6
+15	13.0
+20	17.0
+25	23.1

Temperature = 0°C

Capacity = 4.2 gr/m^3

Amount of Water Vapor in Air = 2.1 gr/m^3

Relative Humidity $= \dfrac{2.1}{4.2} \times 100 = $ 50%

Figure 4–1

The temperature at which air becomes saturated is called the **dew point.** If air cools to a temperature lower than the dew point, the amount of water vapor in the air exceeds the air's capacity to hold it. Hence, the excess water vapor condenses into minute liquid water droplets on solid surfaces. Such solid surfaces are essential for condensation, because minute water droplets are too small and inactive to combine sufficiently to form droplets large enough to be visible; thus, solid surfaces are a necessary vehicle. Fortunately, there are always sufficient microscopic dust and salt particles floating about in the atmosphere to provide the necessary surfaces for condensation. When water vapor begins to condense on these particles, they become what are termed **condensation nuclei.** Some condensation nuclei are especially well adapted to attracting liquid droplets; they are called *hygroscopic nuclei.*

Temperatures generally decline with an increase in altitude in the troposphere (see Module 2). If air rises, therefore, it will cool; and if it cools sufficiently, it may reach the dew point, where the condensation process can begin. The elevation at which the dew point is reached is called the **condensation level.**

When air moves vertically in the atmosphere, its temperature changes even though no additional calories of heat energy are added to or subtracted from the air. The temperature changes only because of the change in *volume* of the air. As air rises or sinks in the atmosphere, it will encounter lower or higher pressure, respectively. As air rises, it expands because air pressures decrease with height. When air expands, it cools because the heat energy in the air becomes less concentrated. Conversely, if air sinks, its volume contracts due to increasing pressure, and, in turn, temperature rises due to the increased concentration of heat. Temperature changes due to expansion and contraction of air are called **adiabatic temperature changes.**

The rate at which air cools or warms adiabatically depends on whether the air is saturated or unsaturated. The rate of adiabatic temperature change in dry or unsaturated air (less than 100% relative humidity) is a constant 1°C/100 m (5.5°F/1000 ft). This rate of temperature change with altitude is called the *dry adiabatic lapse rate.* As unsaturated air rises, it cools at the dry adiabatic rate until it reaches the dew point and condensation level, where it is saturated (100% relative humidity). Once the dew point is reached, the condensation process begins. As a result, if the air continues to rise, it will cool at a new rate called the *wet adiabatic lapse rate.*

The wet adiabatic rate differs from the dry adiabatic rate because as water vapor condenses, latent heat of condensation is released to the air and raises the air temperature. In effect, two opposing processes are occurring simultaneously: (1) air is cooling by expanding as it rises and (2) air is warming from the addition of latent heat of condensation. The net result of these two processes is that air continues to cool, but it cools at a slower rate. When no heat is added to the air, as with the dry adiabatic rate, the air cools at 1°C/100 m.

With the addition of heat to the air, however, as with the wet adiabatic rate, the air cools at some rate less than 1°C/100 m. The wet adiabatic rate, for example, may be 0.5°C/100 m. The wet adiabatic rate is not a constant rate; rather, it varies with air temperature. Since warm air is capable of holding more water vapor than is cold air, the amount of water vapor condensed and the amount of latent heat of condensation released are greater in warm air. Consequently, the wet adiabatic rate is less in warm air than in cold air.

Precipitation

As water vapor condenses in the atmosphere, the minute liquid droplets collect on condensation nuclei to eventually form clouds. It takes many millions of droplets to form just one cloud. It requires additional mechanisms for cloud droplets to combine into larger drops heavy enough to fall to the ground as precipitation. If these mechanisms are not active, clouds may form in abundance but yield no precipitation.

Precipitation is a general term for any type of moisture that falls from the atmosphere to the earth's surface. All forms of precipitation—rain, snow, sleet, and hail—require the combination of many cloud droplets into larger drops. Two mechanisms act to combine cloud droplets into larger drops: (1) the Bergeron process and (2) the collision-coalescence process.

The *Bergeron process* (Figure 4–2) depends on the presence of freezing nuclei to attract cloud droplets to them and form tiny ice particles. The ice particles, in turn, more actively attract additional cloud droplets than do the water droplets. As a result, the ice particles grow rapidly at the expense of the water droplets. Eventually, the ice crystals grow too large to remain suspended in a cloud, and as they fall to the ground, they melt into raindrops. Cloud seeding uses the same principle as the Bergeron process: Silver iodide crystals are released into a cloud, where they act as freezing nuclei, actively attracting cloud droplets to form raindrops.

The *collison-coalescence process* functions as a result of the varying size of suspended water droplets in clouds. Large droplets fall at a faster rate within a cloud than do small droplets. As the large droplets fall, they collide with smaller droplets and coalesce with them into a single larger drop. The combination of a large number of these cloud droplets can eventually produce a raindrop. It is most likely that both the Bergeron and the collision-coalescence processes work in combination much of the time.

Snow is the most common form of solid precipitation. It forms at temperatures below freezing where water vapor **sublimates** (converts from a gas to a solid) into ice crystals. Several ice crystals combine to form a snowflake. It requires many fewer ice crystals to form a snowflake than it does cloud

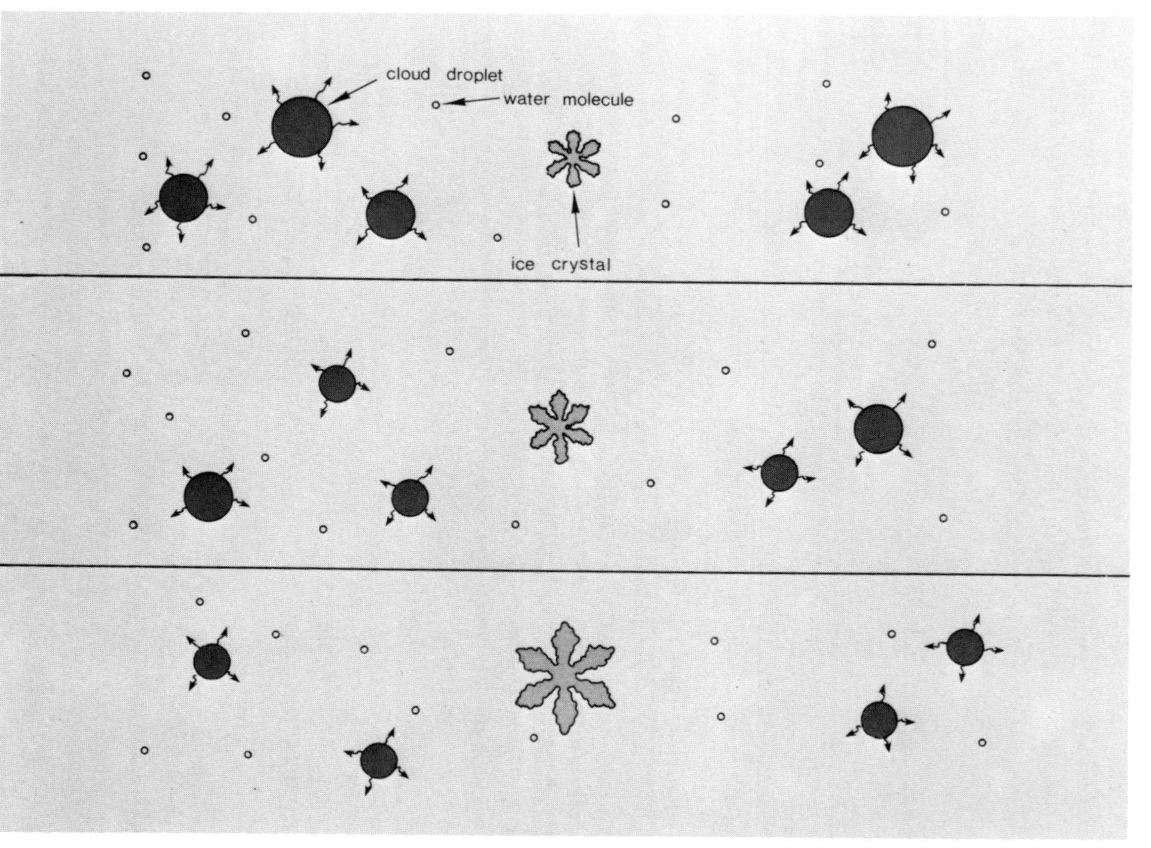

Figure 4–2

droplets to form a raindrop. Consequently, it takes, on the average, 10 inches of snow to equal 1 inch of water.

Sleet is simply partially frozen raindrops that form as raindrops fall through a freezing layer of air near the earth's surface. Hailstones are chunks of ice that form around freezing nuclei in a cloud. They form almost exclusively in well-developed cumulonimbus clouds, and they remain suspended and grow due to the strong updrafts characteristic of these clouds. Hailstones generally rise and fall within a cumulonimbus cloud several times before they become too heavy to remain suspended. As a result, each time the hailstone is lifted, it attracts more moisture and grows in size.

Precipitation Methods

Large masses of air must be lifted and cooled sufficiently for condensation and precipitation to develop. There are three basic methods by which large masses of air rise vertically in the atmosphere: (1) by convection, (2) by orographic lifting, and (3) by cyclonic circulation.

First, in the *convection process,* air near the surface is heated, becomes less dense, and, as a result, rises. If it rises and cools sufficiently to reach the dew point, condensation will occur and precipitation may follow. Convectional precipitation, which is most common during hot summer days, was thought to be associated with single cumulus and cumulonimbus clouds that developed over local ''hot'' spots at the surface. More recent investigations support the idea that these clouds are associated with larger-scale circulations of air and organized systems of clouds rather than single, isolated individuals. Convectional precipitation is most often characterized by brief, heavy downpours that are highly localized.

Second, air rises *orographically* as it runs into major physical barriers, such as steep coastlines or mountains. As air flows up a mountain slope, it cools and may reach the dew point. Condensation and precipitation may then develop. The slope of a mountain facing into the wind is called the **windward slope;** it is this slope that receives most of the precipitation. The slope facing opposite the wind is called the **leeward slope.** This slope is usually typified by arid conditions since the air circulation on this slope is sinking and therefore the air is being warmed, activating the evaporation process. Orographic precipitation may be in almost any form and in almost every degree of intensity.

The third basic method for producing precipitation is by *cyclonic circulation.* The circulation of air in a cyclone, or low-pressure cell, usually involves two unlike masses of air: a cold air mass and a warm air mass. These air masses come into contact and form warm and cold fronts. Due to differences in density, when two unlike masses of air come into contact, the less dense air (warmer air) is forced to rise above the colder air. Thus, it is along warm and cold fronts that warm air is forced upward in a cyclone. Lifted masses of warm air may cool sufficiently to develop frontal precipitation. Cyclonic precipitation, or frontal precipitation, is the most common type of moisture-producing activity in the middle latitudes where warm and cold air masses are constantly drawn into conflict. (A more detailed description of middle-latitude cyclones follows in Module 5.)

The distribution of precipitation in the world forms a variable pattern, but generally speaking, average annual precipitation amounts are greatest in the world low-pressure belts and on land locations near coastlines (see Figure 4–3). Such average figures of precipitation provide, however, no more than a general guide to moisture expectations. More critical from a human standpoint is the *degree of departure* from average amounts. Locations that generally receive more than 100 inches of rainfall annually are likely not to suffer water shortages even if the total for any one year were to drop as much as 10 inches or more. But for those people who live in regions where the average annual totals are only about 20 inches, a decrease of 10 inches could have disastrous results. Unfortunately, it tends to be true that the regions that are deficient in moisture under normal circumstances are the same regions where departures from normal are most common.

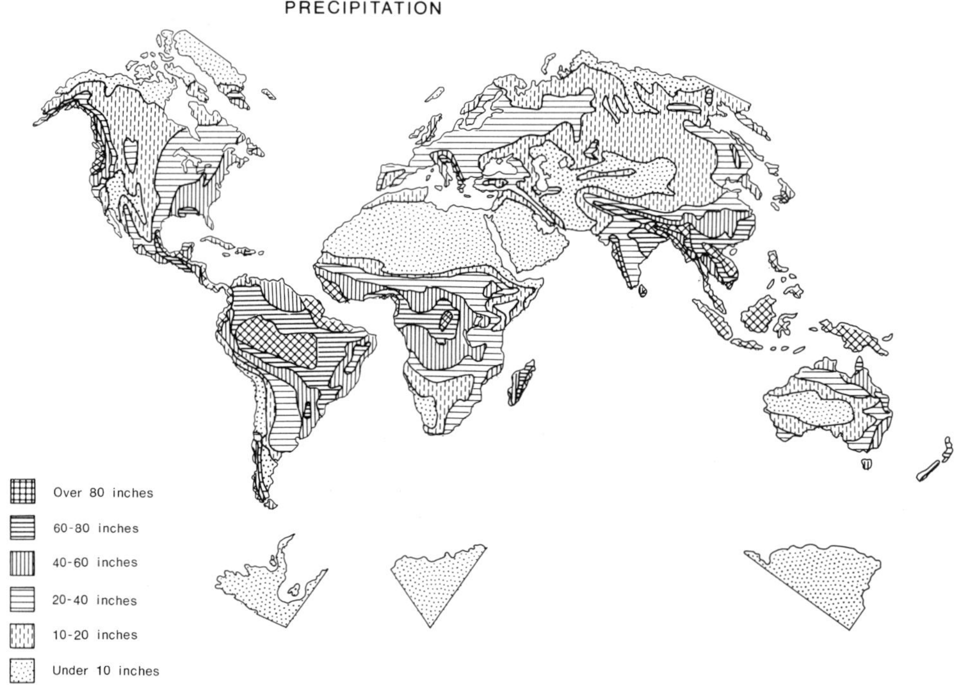

Figure 4–3

Deviations from annual amounts are important, but seasonal deviations can be of even greater significance. Summer is the growing season for most crops in the middle latitudes, and it is also the season of most abundant precipitation. Summer droughts, as experienced in the summer of 1976 in the central sections of the United States, resulted in economic disaster for many farmers. In addition, most of central and southern California failed to receive the normal amount of precipitation during the winter of 1976–77. As a result, drought plagued those areas during the following summer, resulting in water rationing and crop failures in many areas. When one is analyzing total precipitation amounts, it is important to realize that so-called normal or average amounts are more often the exception than the rule.

Additional Material

Clouds

Clouds may form only a few inches above the earth's surface if the air is so highly saturated that it needs to be lifted only a few feet to be adequately cooled. Or clouds can form near the surface if they are cooled by conduction by the cold ground beneath the air. Clouds near or at ground level are called *fog*.

Most clouds, however, form high in the atmosphere where temperatures often are below freezing level. Below freezing, water vapor may sublimate directly into ice crystals. Some clouds contain only water droplets, while others may be composed entirely of ice crystals, and still others may contain a mixture of both. In fact, clouds are often classified by their composition, as water clouds, ice clouds, or mixed clouds.

Water clouds are generally easy to distinguish from ice clouds by their appearance. They possess a sharply defined outer margin, whereas ice clouds most often appear fibrous without a distinct border. Also, water clouds in the middle and higher troposphere will ofter cast shadows on clouds that have formed at lower altitudes. Ice clouds, on the other hand, almost never cast shadows. If one looks at the sun through a water cloud, a dull-colored circle appears to outline the sun with bright outer edges. This appearance is termed a *corona*. Ice clouds, on the other hand, will appear to form a halo effect around the sun which will be larger and brighter than the corona.

Another common method for classifying clouds is by their height above the ground. Although this method is useful for routine observations, it is open to a large degree of subjectivity. There are three main families of clouds based on altitude: low clouds, middle clouds, and high clouds. For their convenience, meteorologists have set the approximate limits for each category as follows: 0–2500 meters (8250 ft) for low clouds; 2501–6000 meters (19,800 ft) for middle clouds; and 6001+ meters for high clouds. A fourth family based on altitude is the towering clouds, which have their base at the low-cloud level but which extend upward through all the altitude classes.

Clouds form when a sufficiently large and dense body of air has been lifted vertically and cooled sufficiently for the water vapor in the air to condense. Clouds will assume definite shapes based on the degree of vertical lifting of the air. If a mass of air is being forced to rise rapidly, the clouds will generally assume a great deal of vertical height. These puffy-shaped clouds are said to have a *cumulus* shape. On the other hand, if a mass of air is rising very slowly, the clouds will have little vertical thickness and will appear layered or sheetlike. These clouds are said to have a *stratus* shape.

As you learned previously, air that rises vertically without much effort is said to be *unstable*. Stable air, on the other hand, is air that generally resists vertical motion. Clouds that appear puffy and often occur as separate individual clumps normally develop under the influence of warm temperatures and unstable conditions. Clouds that appear as unified, gray sheets or layers normally develop under very mild, unstable conditions. Keep in mind, of course, that the atmosphere is constantly in motion. Thus, clouds may be forming under mildly unstable conditions one moment and suddenly change due to a change in conditions.

Adiabatic Lapse Rates

Adiabatic lapse rates apply when air rises or sinks. Figure 4–4 is an idealized diagram that demonstrates the application of adiabatic lapse rates as air is forced to rise orographically over a mountain. The diagram assumes that the air approaching the mountain at point A is unsaturated (relative humidity is less than 100%). As unsaturated air begins to ascend the windward slope, the air temperature declines at the dry adiabatic rate (1°C/100 m). The air eventually cools sufficiently to reach the dew point (100% relative humidity). The condensation level is assumed to be at point B.

The air is now saturated and condensation begins. If the air continues to rise from point B to point C, temperatures will continue to decline and the water vapor in the air will condense. Consequently, due to the release of latent heat of condensation, the air cools at a slower rate (the wet adiabatic rate) from point B to point C.

Once the air reaches point C at the peak of the mountain, it begins to flow down the leeward side of the mountain toward point D. As the air sinks, it is compressed and the temperature rises. Consequently, the air temperature is immediately higher than the dew point, the air is no longer saturated, and the condensation process ceases. From point C, and as long as the air continues to sink, the dry adiabatic lapse rate applies; in this case, the temperature rises at the rate of 1°C/100 m. In fact, because the air temperature rises from point C to point D, the air's capacity to hold water vapor increases, and the potential evaporation increases.

ADIABATIC LAPSE RATES

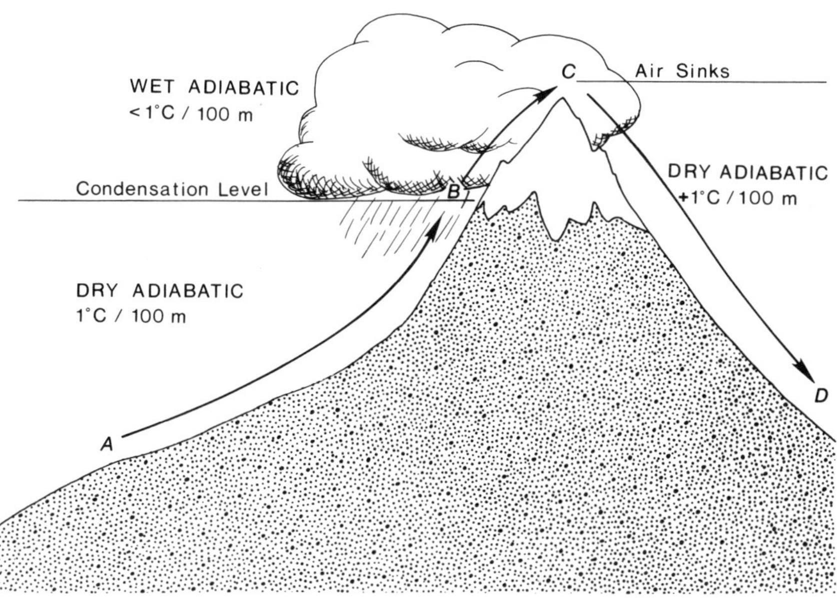

Figure 4–4

Questions

1. What are the important functions of humidity in the atmosphere?

2. Why is evaporation a cooling process?

3. How does latent heat of vaporization differ from latent heat of condensation?

4. Why does the potential for evaporation increase as temperature increases?

5. Why is absolute humidity rarely used by meteorologists as a measure of water vapor content?

6. Why does moisture condense on the inside of windows in the winter months?

7. Why is the water content of a snowflake generally less than the water content of a raindrop?

8. If unsaturated air rises, it cools at the dry adiabatic lapse rate until its temperature reaches the dew point. What will happen to the temperature of a parcel of air if the parcel is sinking and the water vapor content remains constant?

9. Why are the dry adiabatic and the wet adiabatic lapse rates different?

10. Why do cloud droplets not fall to the ground?

11. How do hailstones form?

12. Why is the leeward slope of a mountain range usually drier than the windward slope?

13. Would you expect to receive more precipitation from a low-pressure or a high-pressure system? Why?

14. In which of the world pressure belts would you expect total precipitation to exceed total evaporation? In which pressure belts would you expect the opposite to be true?

15. Why do most coastal regions receive more total precipitation than most continental interiors?

16. Why might the seasonal distribution of precipitation in a given region be more important than the average annual amount?

17. Why might you feel chilly in a room with a temperature of 72°F and a relative humidity of 20%, but comfortable at the same temperature with the relative humidity at 45%?

Suggested Readings

ANTHES, R. et al. 1975. *The atmosphere*. Columbus, OH: Charles E. Merrill. Chapter 8.

BATTAN, L. 1974. *Weather*. Englewood Cliffs, NJ: Prentice-Hall. Chapter 5.

EARNEY, F. C. F., and KNOWLES, B. A. 1974. Urban snow hazard: Marquette, Michigan. In *Natural Hazards,* by G. White. New York: Oxford University Press. Pages 167–74.

KIRKBY, A. V. 1974. Individual and community response to rainfall variability in Oaxaca, Mexico. In *Natural Hazards,* by G. White. New York: Oxford University Press. Pages 119–28.

LEHR, P.; BURNETT, R. W.; and ZIM, H. S. 1975. *Weather*. New York: Golden Press. Pages 21–33.

MILLER, A., and THOMPSON, J. C. 1975. *Elements of meteorology*. 2d ed. Columbus, OH: Charles E. Merrill. Pages 30–43.

MOLINE, N. T. 1974. Perception research and local planning: floods on the Rock River, Illinois. In *Natural Hazards,* by G. White. New York: Oxford University Press. Pages 52–59.

OLIVER, J. E. 1977. *Perspectives on applied physical geography.* North Scituate, MA: Duxbury Press. Chapters 5 and 9.

TREWARTHA, G. T. 1968. *An introduction to climate.* 4th ed. New York: McGraw-Hill. Chapter 4.

Module 5

Atmospheric Circulation

INTRODUCTION

The air that comprises the atmosphere is constantly in motion due to the earth's rotation and the differential heating of the earth's surface. This circulation of the atmosphere is extremely important in the distribution of heat and moisture from place to place. Without this circulation of heat and moisture, many places on the earth would be too cold, too hot, or too dry to be habitable. The atmosphere transports heat from equatorial regions to polar regions and moisture from regions of excess to regions of deficiency. About 80% of the heat redistribution over the earth is accomplished by atmospheric circulation, with the remainder of the heat redistributed by ocean currents.

In addition to equalizing heat and moisture over the earth's surface, air circulation also transports pollutants many thousands of miles from their source and determines the direction of flow of weather systems. It also affects the daily operation of ship and air transportation. Thus, knowledge of atmospheric circulation is quite important.

Within the general atmospheric circulation is the complex interaction between surface weather systems and upper-air weather, which can interact to cause extreme weather conditions such as prolonged droughts, excessive rainfall, thunderstorms, tornadoes, and hurricanes. To improve our weather knowledge, meteorologists are trying to monitor as much of the atmosphere's circulation as possible around the world. International programs such as World Weather Watch (WWW) and the Global Atmospheric Research Program (GARP) have been launched in a major attempt to constantly monitor the state of the atmosphere for almost any place on earth.

The various air masses that invade North America are part of the atmosphere's circulation. As the masses invade, they bring with them a variety of weather characteristics that range from cold and dry continental polar air masses to warm and moist maritime tropical air masses. In the middle latitudes, it is the continual conflict between contrasting air masses that establishes weather fronts and frontal weather activity. Frontal weather activity, which is commonplace in the middle latitudes, is much rarer at tropical and polar latitudes where conflicts between air masses are much less frequent. Thus, the atmosphere circulates around the earth and contains within it a wide variety of subsystems: the world surface-wind systems, upper-air winds, air masses, the evaporation-precipitation system, and active weather fronts. Because of the many interrelated systems that are closely connected to air circulation, it is of vital importance to understand the general nature of atmospheric circulation.

OBJECTIVES

By the end of this module, you should be able to do the following:

1. Describe the derivation and general weather characteristics of the four principal types of air masses.
2. Describe the weather characteristics of the air masses that invade North America.
3. Compare and contrast surface-air circulation with upper-air circulation.
4. Describe the relationship between surface-air circulation and the jet streams.
5. Describe how surface weather is affected by the direction of the jet stream.

6. Describe the life cycle of a middle-latitude cyclone.
7. Describe the nature of the weather associated with cold fronts, warm fronts, and occluded fronts.
8. Describe the general characteristics of a mature middle-latitude cyclone in terms of the location of fronts; the locations of the warm, cold, and cool sectors; and the general precipitation pattern.

KEY TERMS

air mass   front

source region   cyclogenesis

jet stream   occluded condition

Now you are ready to begin the audiovisual portion of this module. Select the MEDIAPAK 5 components(s) and proceed. Following is a topical outline of the audiovisual sequence. You will find this outline helpful for reference and review. After completing MEDIAPAK 5, return to this book to perform the exercises.

OUTLINE

Air masses
> *Source regions*
> *Characteristics*
> *North American air masses*

Polar front

Upper-air circulation
> *Jet stream*

Middle-latitude cyclone
> *Life cycle*
> *Frontal weather*
> *Circulation of a cyclonic system*

Exercises

1. Which of the following air masses will most likely have its source region over a land surface at high latitude?
   A. mP
   B. mT
   C. cP
   D. cT

2. Which of the following North American air masses is warm, moist, and unstable all year and invades the continent east of the Rocky Mountains?
   A. cP
   B. mP (Atlantic)
   C. cT
   D. mT (Gulf)
   E. mT (Pacific)

3. Which of the following North American air masses becomes highly stable in the summer and is responsible for dry summers along the west coast between 25° and 55° latitude?
   A. cP
   B. mT (Pacific)
   C. mP (Pacific)
   D. cT
   E. mT (Gulf)

4. The polar front is located between 30° and 60° latitude. It is farthest poleward in
   A. summer.
   B. winter.

5. The probability of active cyclonic formation increases when the direction of flow of the upper westerlies is generally from _____________ to _____________.
   A. east; west
   B. west; east
   C. south; north
   D. north; south

6. The initial identification of a cold and warm front within a middle-latitude cyclone is typical of the _____________________________ stage of its life cycle.
   A. open
   B. occluded
   C. formation
   D. dissipating

7. Within a middle-latitude cyclone, a warm front advances at a more rapid rate than does a cold front.
   A. True
   B. False

The following diagram shows a mature middle-latitude cyclone in the Northern Hemisphere. Refer to this diagram when answering Exercises 8, 9, and 10.

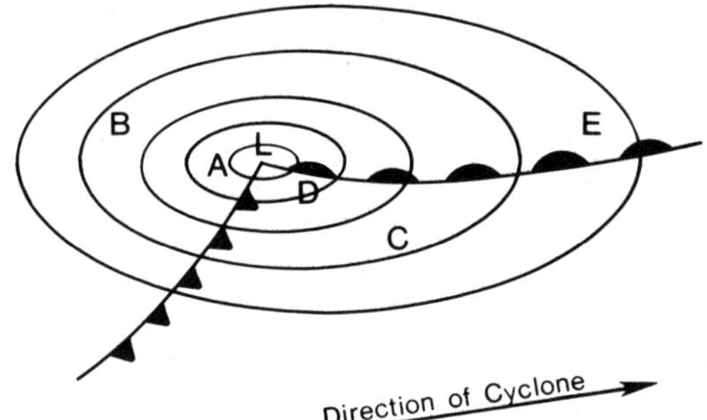

8. Which letter lies within the warm sector of the system?
   A. *A*
   B. *B*
   C. *C*
   D. *D*
   E. *E*

9. What is the most likely wind direction at letter *B?*
   A. Northeast
   B. Southwest
   C. Southeast
   D. Northwest
   E. South

10. At which of the following letters will the barometer be rising?
    A. *A* and *B*
    B. *A, E,* and *D*
    C. *B* and *C*
    D. *B, E,* and *A*
    E. *C, E,* and *D*

11. During the summer months, the chances for severe weather are generally greater
    in your area if the center of a middle-latitude cyclone passes
    _________________________ of you.
    A. poleward
    B. equatorward

Summary

The air in the atmosphere is constantly in motion. **Air masses,** large bodies of
air many thousands of square miles in size, move from points of origin to
other areas, often drastically changing the weather in the new location. These
air masses are defined by their relatively uniform temperature and moisture

characteristics; for example, an air mass may be warm and wet, cold and dry, and so forth. The characteristics of an air mass are derived from the general nature of the **source region,** the environment above which the air mass originates. For example, if a tropical ocean surface is the source region for an air mass, that air mass will likely be characteristically warm and abundant with moisture. Not all places on the earth are potential source regions for air masses due to the nature of the earth's surface or the circulation of the atmosphere in a particular location. To be a source region, a location must have relatively uniform surface characteristics and must be large enough to underlie one day's journey of air.

Source regions for air masses in North America include the subarctic plains of northern Canada, the southwestern deserts, the Gulf of Mexico, the Pacific Ocean off the coast of British Columbia and the Baja Peninsula, and the Atlantic Ocean off the coast of New Foundland and the Carolinas. The characteristics of source regions vary with season, and so also do the temperature and moisture properties of the air masses that originate there.

Most source regions are also regions where high pressure dominates. Low-pressure regions are not conducive to air mass formation because they tend to draw in contrasting types of air, for air masses flow from their source regions toward regions of lower pressure. As air masses move into regions unlike their source regions, they can generate major changes in the daily weather at those newly invaded areas. For example, a middle-latitude area might have its weather dominated by a warm, moist tropical air mass one day and a cold, dry polar air mass the next. The greater the distance an air mass moves from its source region, the more the air mass will be modified by its new environment. Distances of thousands of miles may be traversed, however, with only a slight modification of the properties of an air mass.

Air Mass Classification

There are four basic types of air masses: (1) warm and moist, (2) warm and dry, (3) cold and moist, and (4) cold and dry. Warm air masses originate at tropical or subtropical latitudes, and cold air masses at polar regions. Air masses that have source regions over water surfaces are usually moist and are called *maritime* air masses. Dry air masses have source regions over land areas and are called *continental* air masses. By combining these characteristics, meteorologists have defined four classes of air masses:

1. maritime Tropical
2. maritime Polar
3. continental Tropical
4. continental Polar

On weather maps, air masses are symbolized with letter designations. A lowercase *m* means *maritime;* a lowercase *c* means *continental;* a capital *T* means *tropical;* and a capital *P* means *polar*. The four classes of air masses in symbol form are

1. maritime Tropical (mT)
2. maritime Polar (mP)
3. continental Tropical (cT)
4. continental Polar (cP)

In addition, air masses can be further classified by the relationship between the temperature of the air mass and the surface temperature beneath it. A capital *W* indicates that the air mass is warmer than the underlying surface, and a capital *K* indicates that the air mass is colder than the surface temperature. Air masses with a K designation have a higher probability of producing precipitation because the lower layers of the air mass are being warmed, which promotes a vertical lifting of the air.

North American Air Masses

The air masses that invade North America are summarized in Table 5–1.

Fronts

As an air mass invades new territory, it often meets a second air mass with differing temperature and moisture properties. Such a boundary between two unlike masses of air is called a **front.** It is along fronts that most active weather forms, at least in the middle latitudes. The greater the degree of contrast across a frontal boundary, the greater is the potential for active weather to develop. In fact, the entire concept of fronts and frontal disturbances is almost exclusively a feature of the middle latitudes, which receive continual invasions of contrasting polar and tropical air masses.

A major feature of middle-latitude weather is the polar front, the boundary between cold polar air masses and warm tropical air masses. The polar front is a nearly constant boundary that varies in location between 30° and 60° latitude during the year. It is farthest equatorward in winter and shifts poleward in summer. Most of the active weather in the middle latitudes is closely associated with the polar front.

The boundary of the polar front is wavelike in appearance, with the crest of the waves pointing poleward (Figure 5–1). Some of these waves develop into active cyclones often associated with upper-air circulation. Winds in the

Table 5–1. *Major North American Air Masses.*

Class	Source Region	Characteristics	Major Regions of Influence
Continental polar (cP)	Arctic plains of northernmost Canada.	*Winter:* Very cold, dry air. May bring record cold temperatures. *Summer:* Poorly developed. Clear.	Eastward of Rocky Mountains, flows as far south as Oklahoma in winter. Remains close to source in summer.
Continental tropical (cT)	Deserts of southwestern United States and northern Mexico.	*Winter:* Does not develop. *Summer:* Very hot and dry.	Southwestern states with occasional outbreaks into the central Great Plains.
Maritime polar, Atlantic [mP (Atlantic)]	North Atlantic Ocean off coast of New Foundland.	*Winter:* Rarely invades North America. Freezing temperatures, variable amounts of precipitation (damp). *Summer:* Clear skies, moderate temperatures.	Northeastern coast from Cape Hatteras to Baffin Island and east of Appalachians.
Maritime polar, Pacific [mP (Pacific)]	Northern Pacific off coast of British Columbia.	*Winter:* Warm and humid. Generates showers, fog, snow in higher elevations, mild temperatues in central North America. *Summer:* Cool, relatively dry weather.	West coast from central California to Aleutians. Greatly modified east of Rockies in central sections of continent.
Maritime tropical, Gulf [mT (Gulf)]	Gulf of Mexico, Caribbean Sea, subtropical Atlantic Ocean.	*Winter:* Moderate temperatures and showers. May provide moisture source for snow in Midwest. *Summer:* Warm and humid, heavy rains.	Dominates in summer east of Rockies north to southern Canada. In winter affects mainly southeastern states.
Maritime tropical, Pacific [mT (Pacific)]	Pacific Ocean off coast of Baja Peninsula.	*Winter:* Fog and precipitation, moderate temperatures. *Summer:* Warm and dry.	Affects mainly northwestern Mexico and southwestern United States. May affect areas north to Washington but of little importance east of coastal mountains.

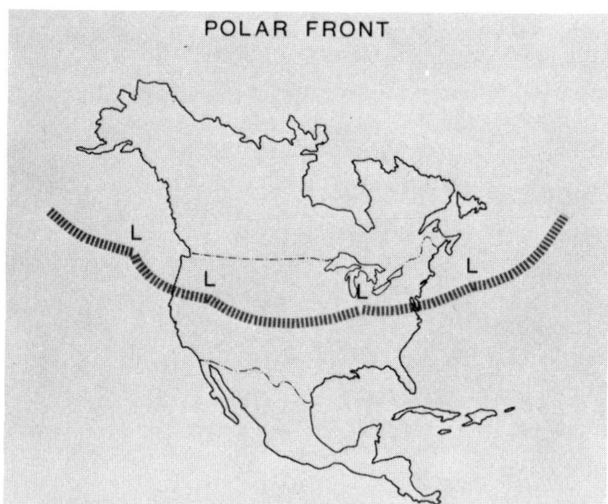

Figure 5–1

upper troposphere have much higher velocities than do winds near the surface, due to larger pressure gradients and lack of friction. Indeed, wind velocities average 39 knots (47 mi/hr) in summer and 78 knots (94 mi/hr) in winter.

The upper-air wind belts consist of a relatively weak easterly flow over polar latitudes, a better-developed easterly flow over the tropics, and a very well developed westerly system between 20° and 75° latitude. The upper-air westerlies, in fact, are best developed in about the same position as the polar front at the surface. Here, wind velocities may exceed 150 knots (180 mi/hr). The regions of highest wind speeds are referred to as the **jet streams.** The jet stream over the middle latitudes meanders back and forth in a snakelike pattern, completely encircling the earth (Figure 5–2). The number of meanders in the jet stream varies from three to six during the year, with four the most frequently occurring number.

The number of meanders in the jet stream and the path the stream follows are of great importance to weather in the lower troposphere because the direction of air circulation near the surface is dictated largely by the direction of flow of the jet stream. Due to friction, the jet stream and the upper westerlies drag air near the surface along with them. Where the jet stream and the upper westerlies flow equatorward (Figure 5–3), upper air sinks and compresses air at the surface. Conversely, where the jet stream and the upper-westerlies flow poleward, the upper air diverges aloft, triggering a compensating upward flow from the surface which favors the development of low pressure at the surface. Thus, the upper-air circulation pattern can initiate the development of cyclonic and anticyclonic circulations in the lower atmosphere. Whenever meteorologists forecast the weather, not only must they be aware of cyclonic and anticyclonic activity at the surface, but also they must

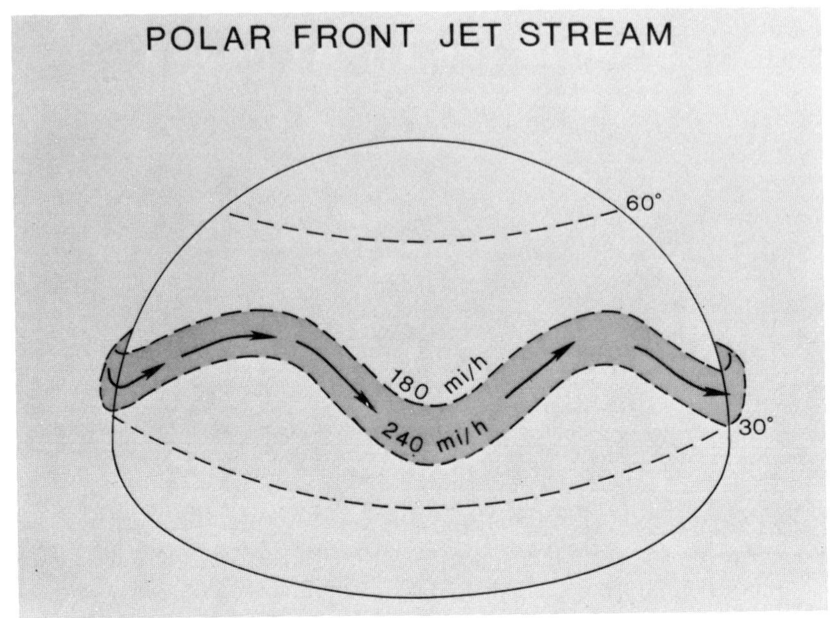

Figure 5–2

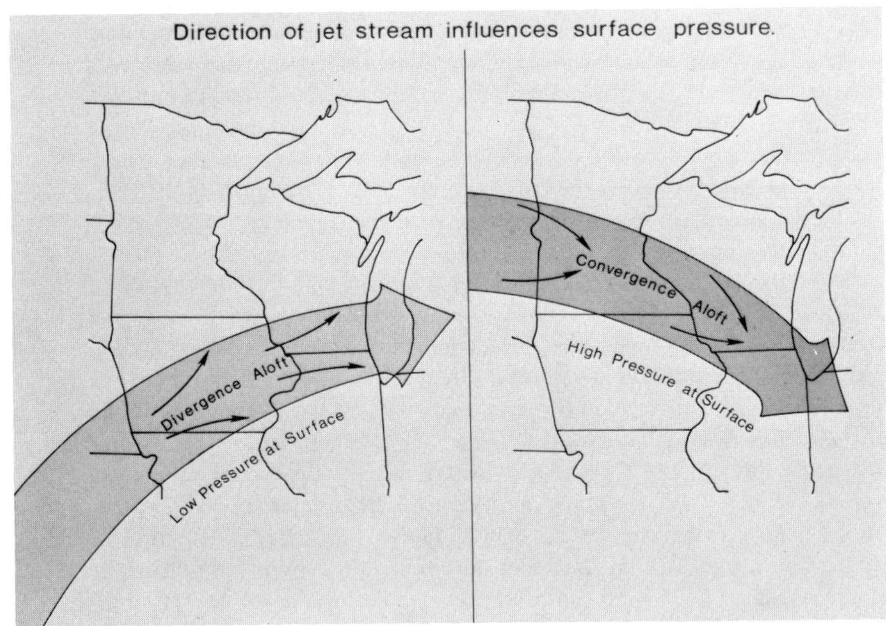

Figure 5–3

analyze the pattern of upper-air circulation which can have a profound influence on surface circulation.

More than one jet stream operates in the atmosphere, but the middle-latitude, or polar-front, jet stream is by far the best developed. The other jet streams include the subtropical jet stream, which is a relatively permanent feature near 30° latitude and is well developed in winter, but very weak and irregular in summer. Two other jet streams are seasonal features in the upper troposphere. The tropical easterly jet stream is in evidence only in summer at latitudes of 10° to 15°. The polar-night jet stream develops only in winter at latitudes poleward of 70°.

Middle-latitude cyclones are best developed along the polar front. They develop as unstable waves along the front triggered by surface pressure variations or upper-air circulation. Once **cyclogenesis,** or birth of a cyclone, begins, the cyclone usually passes through a series of growth stages, growing to maturity and eventually dissipating. This cycle of development is called the *life cycle* of a middle-latitude cyclone.

The cycle begins as in Figure 5–4A, with two unlike air masses moving in opposite directions on either side of the polar front. In the first stage, the formation stage, a wave begins to form in the polar front due to pressure gradients, by air motion on the lee side of mountains, or by upper-air winds. Once the wave has formed, the Coriolis force draws the cold air mass into the warm air mass; a warm front and a cold front are now identified (Figure 5–4B). The two air masses move into the area of the other, much like two people entering a revolving door from opposite sides. The cold air mass has a greater density than the warm air mass, and consequently the cold front advances at a faster rate than does the warm front. After about 36 hours, the cyclone appears as in Figure 5–4C and is now in the second, or mature, stage.

In the mature, or open, stage, a distinct wave has been formed by the frontal boundaries. The warm air is lifted from the surface as it encounters denser, colder air. Condensation and precipitation can form in association with these frontal lines. It is during the mature stage that most of the active weather in the cyclone develops—thunderstorms, tornadoes, heavy showers, and so forth. You should remember that not only is there motion within the cyclone, but also the entire cyclonic system flows in a generally easterly direction across the middle latitudes.

By the third stage, called the **occluded** stage (Figure 5–4D), part of the warm front has been overtaken by the rapidly moving cold front. The warm front that has been overtaken is now defined as an **occluded front.** Eventually, the entire warm front will be occluded. As a result, warm air will no longer be fed into the system and the contrast between the air masses will disappear. The occluded stage marks the beginning of the end of the cyclone. The final, or dissipating, stage results because there is no longer a contrast in the air masses (Figure 5–4E).

STATIONARY POLAR FRONT
Cold Air Mass
POLAR FRONT
Warm Air Mass
A.
FORMATION STAGE
Cold Air Mass
COLD FRONT
WARM FRONT
Warm Air Mass
B.

MATURE (OPEN) STAGE
W
Direction of Cyclone
E
L
Cold Air Mass
Warm Air Mass
C.

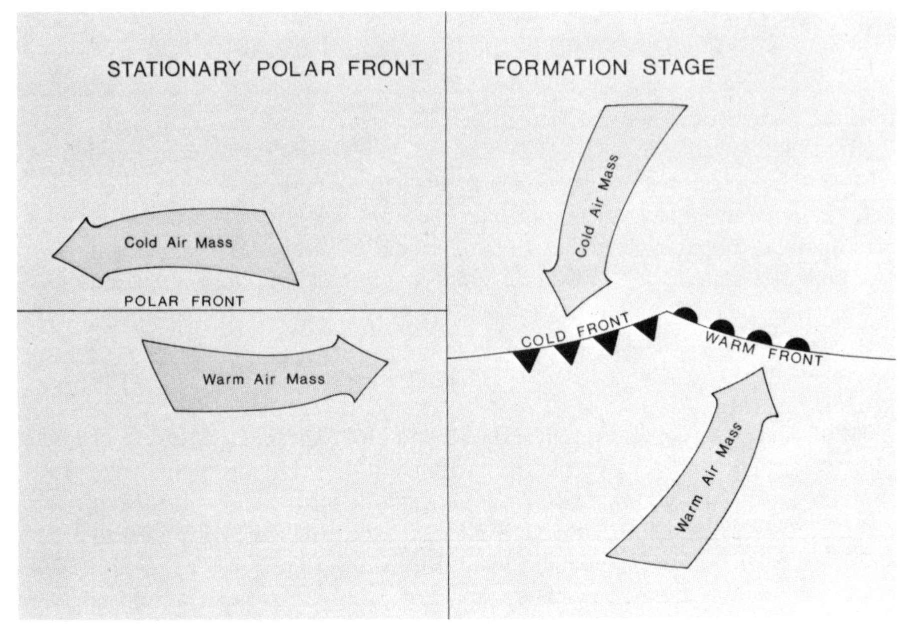

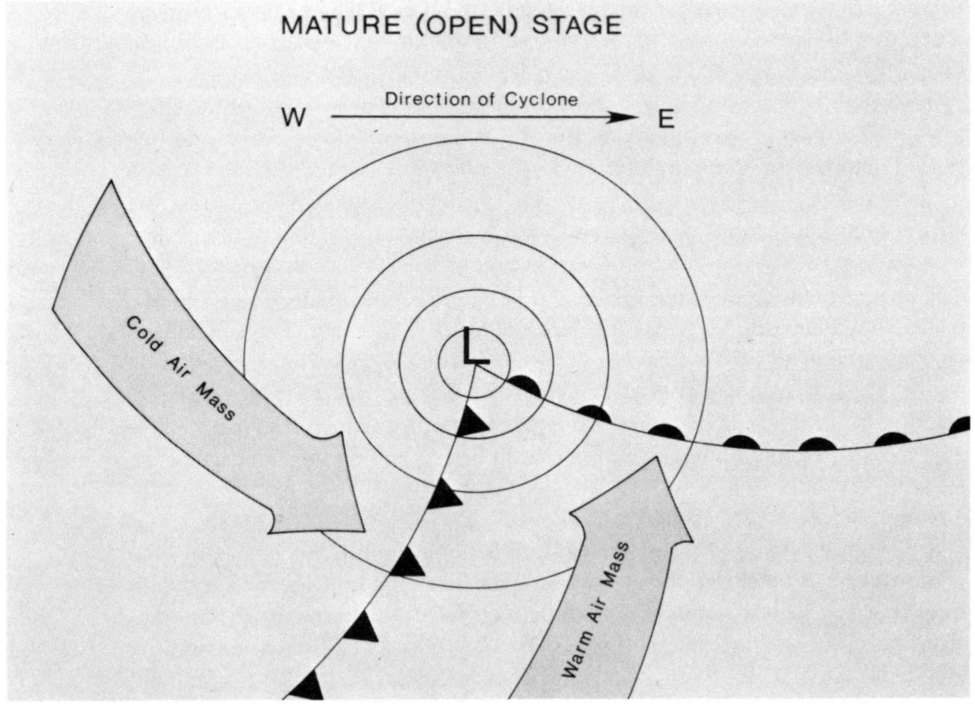

Figure 5–4

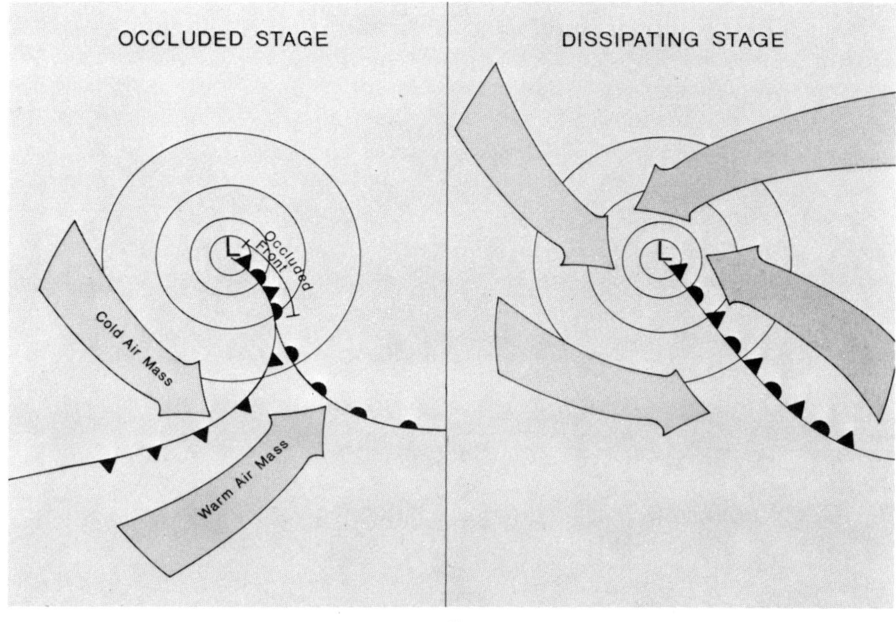

D.   **E.**

Figure 5–4, Continued

Frontal Weather

The active weather in a cyclone is along the frontal boundaries. A cold front marks the leading edge of an advancing cold air mass. As the advancing cold front strikes a warm air mass, the warm air is forced to rise from the surface, and condensation and precipitation often result. As shown in Figure 5–5, a cold front has a relatively steep boundary due to friction with the ground. Cold fronts advance rapidly at rates of 20 to 30 miles per hour. Therefore, due to the size and the speed of the advancing cold front, the warm air is forced to rise very rapidly. Consequently, cold fronts are usually characterized by the rapid buildup of cumulus and cumulonimbus clouds, followed by brief but heavy rainstorms and strong gusty winds. It is generally along cold fronts also that thunderstorms and tornadoes are most common.

The leading edge of an advancing warm air mass marks the warm front (Figure 5–6). As the advancing warm air mass invades a region of cooler air, once again it is the warm air, being less dense, that is forced to rise above a layer of colder air. However, a warm front is considerably less steep than a cold front due to less frictional drag and slower velocity of the front (5 to 15 miles per hour). Thus, layered stratus clouds of varying types build up gradually, and precipitation is typically in the form of light showers, which

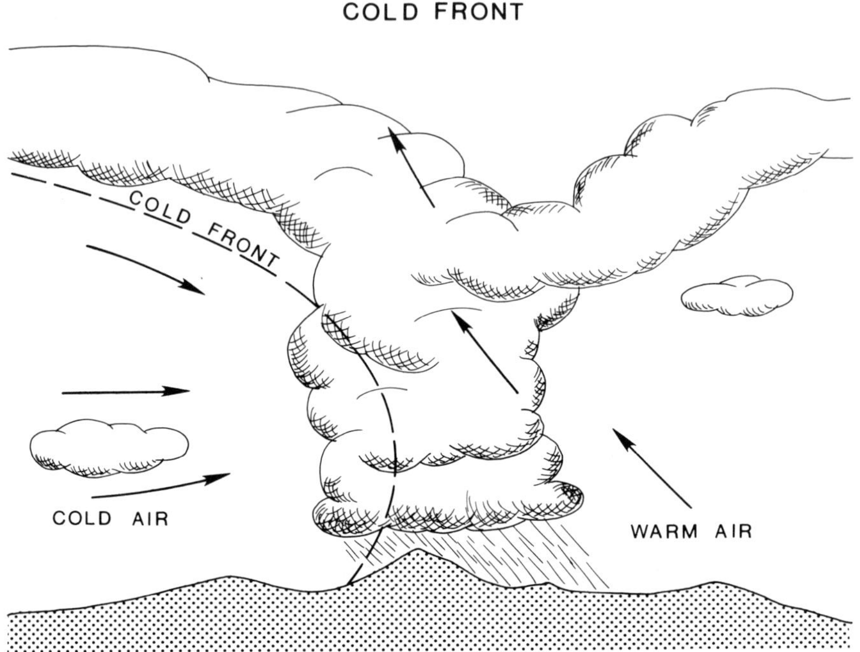

Figure 5–5

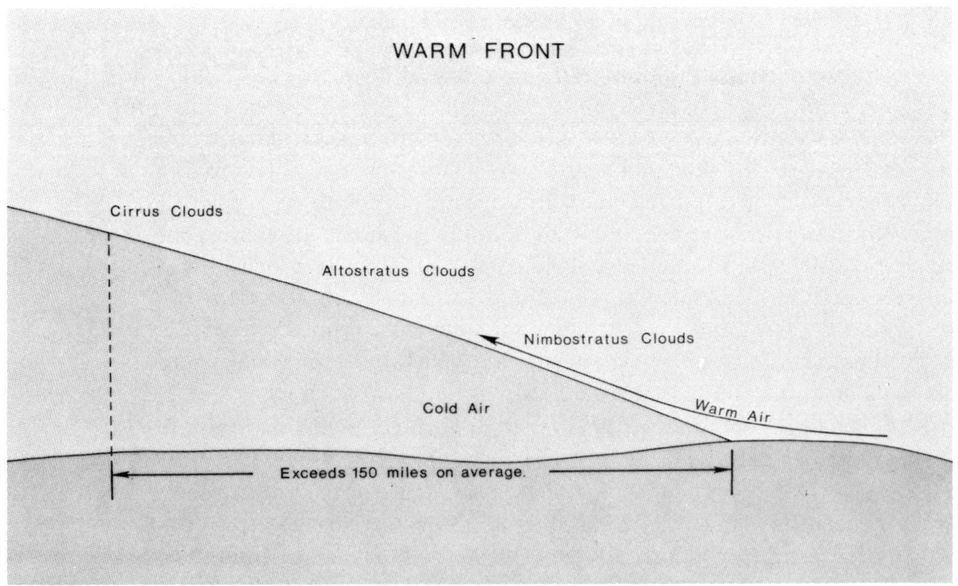

Figure 5–6

can last for many hours from uniformly gray, overcast skies, with light to moderate wind velocities.

The weather associated with an occluded front is most often similar to a modified cold front (Figure 5–7). That is, the warm air is displaced upward from the surface; but since the source of warm air is cut off, clouds do not build a thick sky cover, and precipitation is usually in the form of light showers.

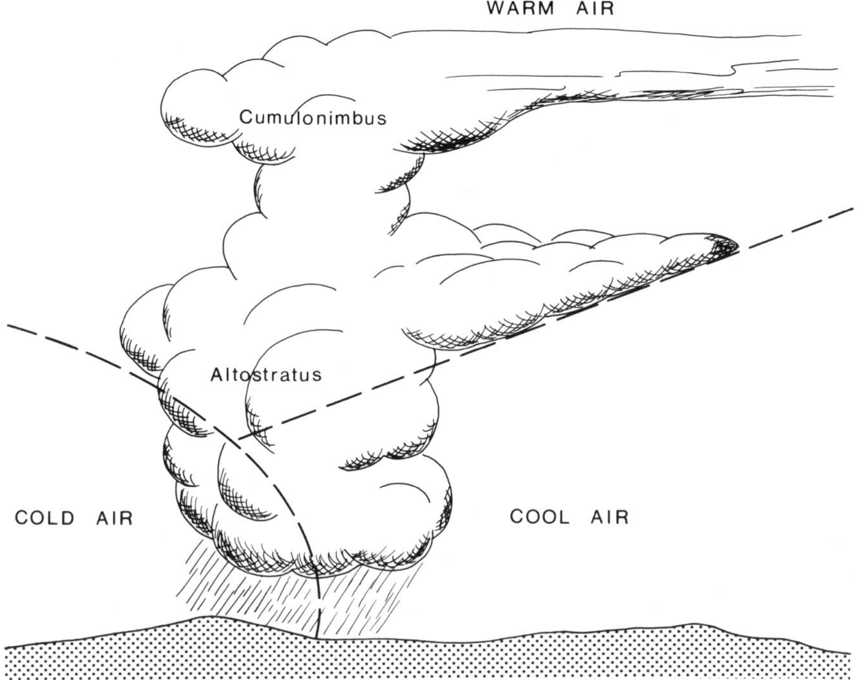

Figure 5–7

In the middle latitudes, cyclones flow in a generally west-to-east direction within the prevailing westerly wind system. As they travel, they tend to follow a definite pattern across the continent. These paths that cyclones follow are called *storm tracks*. The most common arrangement of warm and cold fronts within a cyclone is shown in Figure 5–4C. Because warm fronts generally lie to the east of cold fronts, warm-front weather is the first indication of an approaching cyclone.

Basic knowledge of the arrangement of atmospheric pressure, temperature, precipitation, and winds in a low-pressure cell in the middle-latitudes (Northern Hemisphere) will assist your comprehension of the evening television weather report. Figure 5–8 represents the average pattern of pressure,

MID-LATITUDE CYCLONE

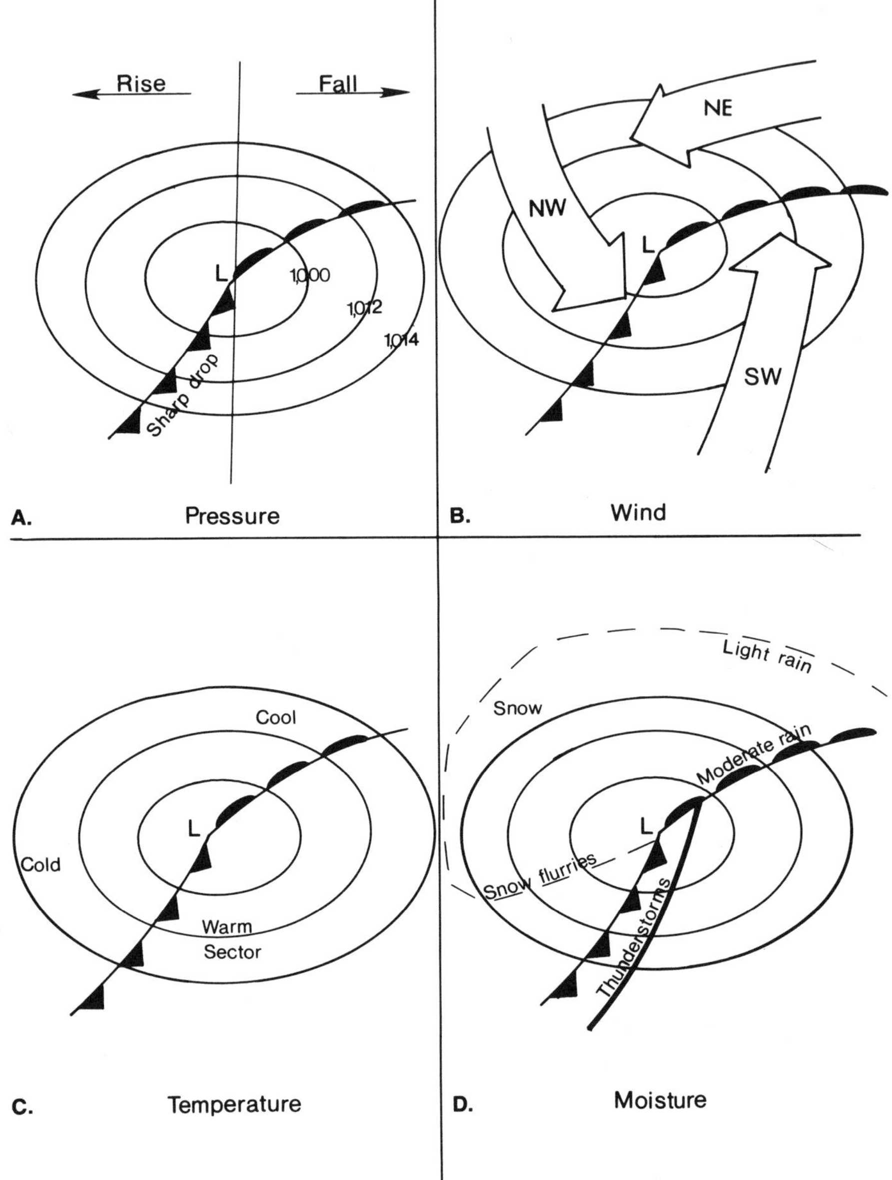

Figure 5–8

temperature, precipitation, and winds around a low-pressure center. If you remember the pattern for each atmospheric variable, you should be able to better understand cyclonic weather.

Figure 5–8A shows the pressure tendency within a cyclone. A falling barometer generally means that a cyclone is approaching. Thus, by watching your home barometer, you can tell if a cyclone is approaching your area. Approaching cyclones bring with them warm and cold fronts and the chance of precipitation. A rising barometer, on the other hand, suggests that the center of the cyclone is moving away from you, to be replaced by an anti-cyclone and fair weather.

Figure 5–8B shows the typical wind circulation around a cyclone. If you know the present wind direction, you should always know where you are relative to the center of the cyclone. Remember, in the Northern Hemisphere, if you turn your back to the wind, your left side will point to the center of a cyclone.

Figure 5–8C shows the general temperature pattern in a cyclone. Of course, actual temperatures vary with the season, and for that reason they are given here only in general terms. Figure 5–8D gives the general pattern of precipitation in a cyclone. It should be apparent that the pattern is directly associated with the location of the fronts within the cyclonic system. It takes an average of three to six days for a cyclone to pass over a given location. In that period of time, the greatest total precipitation is, generally speaking, within a band about 20 to 50 miles wide located about 70 miles north of the center of the low-pressure cell. The heaviest precipitation band is located in that region of the cyclone because that is the area of the most substantial lifting of the warm air mass, which generally contains abundant moisture.

Violent weather is associated with the fronts in a cyclone, particularly the cold front. Therefore, should the center of a cyclone pass poleward of your location, the chances that the cold front will pass through your area are very good. On the other hand, if the center of a cyclone passes equatorward of you, you can expect that the cold front will not push through your area, and the chances of violent weather are considerably less.

Questions

1. What characteristics must an area possess to be an effective source region for air masses?

2. Why do air masses move?

3. What can cause the lower layers of an air mass to become stable or unstable?

4. What air masses in North America regularly invade the region in which you live? During which season is each air mass most likely to appear and what type of weather does each bring with it?

5. What is the significance of the polar front in the middle latitudes?

6. Why do upper-air winds have higher velocities than surface winds?

7. How does the path of the upper-air circulation affect the development of cyclones at the surface?

8. What are the stages of development of a middle-latitude cyclone along the polar front? Describe the characteristics of the cyclone during each stage.

9. How does the weather associated with a cold front compare with the weather associated with a warm front? An occluded front?

10. What sequence of weather changes would you expect if a middle-latitude cyclone center passed just poleward of your location, flowing from west to east? What would you expect if the center passed just equatorward of your region?

Suggested Readings

ANTHES, R. et al. 1975. *The atmosphere*. Columbus, OH: Charles E. Merrill. Chapters 4 and 5.

BATTAN, L. J. 1974. *Weather*. Englewood Cliffs, NJ: Prentice-Hall. Chapters 2, 3, and 4.

LEHR, P. et al. 1975. *Weather*. New York: Golden Press. Pages 60–95.

MILLER, A., and THOMPSON, J. C. 1975. *Elements of meteorology*. 2d ed. Columbus, OH: Charles E. Merrill. Chapters 4 and 5, and pages 147–57.

NEWELL, R. E. 1971. The global circulation of atmospheric pollutants. *Scientific American*, January 1971. Reprint 894. San Franciso: W. H. Freeman.

TREWARTHA, G. T. 1968. *An introduction to climate*. 4th ed. New York: McGraw-Hill. Chapter 5 and pages 193–222.

Module 6

Severe Storms

INTRODUCTION

Storms—thunderstorms, tornadoes, and hurricanes—receive more attention than any other factor in the earth-atmosphere system. People and computers spend countless hours analyzing, trailing, and forecasting atmospheric storms in the attempt to predict where they will strike so that local inhabitants can be forewarned. Each year, thousands of lives are lost around the world due to storms, but forecasting methods that lead to quicker and more accurate warnings are constantly being improved, thereby reducing the probability and occurrence of deaths by storms.

In the United States, the National Weather Service has established an intricate, finely woven network of weather observation stations. National weather data that suggest the possibility of a storm are funneled to the National Severe Storms Forecasting Center in Kansas City, Missouri. The National Center issues proper watches and warnings to those regions likely to be hit by a storm. However, in other countries the system of observation stations and weather computers is often not available. Therefore, efforts to expand these services to all sections of the world must be continued.

Fortunately, severe storms are relatively rare events in any one locality, and, in many cases, they affect only certain sections of a continent or country. Tornadoes, for example, are for the most part a phenomenon of only the middle sections of the United States and almost never develop west of the Rocky Mountains. Hurricanes affect primarily the Gulf Coast and the eastern seaboard and present no threat to the interior of the continent. Thunderstorms are more commonly distributed across the entire country, but they are generally much less devastating than tornadoes and hurricanes.

All severe storms have one common characteristic: They are all a product of the release of latent heat of condensation from abundant water vapor in the air. Thus, storms are most intense when the air is warm and contains large volumes of water vapor, and also when there is a mechanism for rapidly triggering the condensation process.

OBJECTIVES

By the end of this module, you should be able to do the following:

1. Describe both the mechanisms that initiate a thunderstorm and the stages in the life cycle of a thunderstorm.
2. State the distribution of thunderstorms in the United States.
3. Explain lightning and thunder.
4. Describe the weather characteristics and features of tornadoes and waterspouts.
5. Describe the weather characteristics associated with and the main features of a hurricane.

KEY TERMS

storm	tornado
squall	waterspout
lightning	hurricane
thunder	eye

Now you are ready to begin the audiovisual portion of this module. Select the MEDIAPAK 6 *component(s) and proceed. Following is a topical outline of the audiovisual sequence. You will find this outline helpful for reference and review. After completing* MEDIAPAK 6, *return to this book to perform the exercises.*

OUTLINE

Thunderstorms
    Life cycle and characteristics
    Distribution in United States
    Lightning and thunder

Tornadoes
    Formation features
    Distribution in United States
    Waterspouts

Hurricanes
    Formation features
    Weather characteristics

Exercises

1. A thunderstorm in the cumulus stage is characterized by
   - **A.** a squall line.
   - **B.** a downdraft that has spread across the base of the cloud.
   - **C.** warm updrafts of air.
   - **D.** a heavy downpour of rain and hail.
   - **E.** all of the above.

2. The region of strong, gusty winds followed by a heavy downpour is called a(n)
   - **A.** eye.
   - **B.** storm surge.
   - **C.** tornado.
   - **D.** squall.
   - **E.** echo.

3. The region of most frequent thunderstorm activity in the United States is in
   - **A.** Florida.
   - **B.** California.
   - **C.** Ohio.
   - **D.** Oklahoma.
   - **E.** Virginia.

4. Thunder cannot form without lightning.
   - **A.** True
   - **B.** False

5. In the middle latitudes, tornadoes are generally associated with thunderstorms along a(n)
   - **A.** cold front.
   - **B.** warm front.
   - **C.** occluded front.
   - **D.** stationary front.

6. Tornadoes develop most frequently in the
   - **A.** summer.
   - **B.** winter.
   - **C.** spring.
   - **D.** fall.

7. Approximately two-thirds of all tornadoes travel from
   _____________________________ to _____________________________.
   - **A.** northwest; southeast
   - **B.** southwest; northeast
   - **C.** west; east
   - **D.** northwest; southwest
   - **E.** north; south

8. Hurricanes usually intensify when they move over a land surface.
   - **A.** True
   - **B.** False

9. To be classified as a hurricane, a storm must have wind velocities that exceed
   - **A.** 24 mi/hr.
   - **B.** 34 mi/hr.
   - **C.** 54 mi/hr.
   - **D.** 74 mi/hr.
   - **E.** 104 mi/hr.

10. The peak period for hurricanes in the Northern Hemisphere is
   - **A.** June, July, August.
   - **B.** April, May, June.
   - **C.** August, September, October.
   - **D.** October, November, December.
   - **E.** February, March, April.

Summary

Storms are significant departures from normal atmospheric disturbances; they are usually characterized by one or more of the following: strong winds, heavy rains, heavy snows, thunder and lightning. The three severest storms, in terms of the loss of life and property damage that they cause, are thunderstorms, tornadoes, and hurricanes.

Thunderstorms

Thunderstorms—rain showers accompanied by thunder and lightning—are the most common severe storms. It is estimated that between 40,000 and 50,000 thunderstorms develop daily around the earth. A thunderstorm usually will not develop until clouds extend a minimum of 20,000 feet (6 km) above the ground. Some thunderstorm clouds extend into the tropopause at elevations up to 10 miles (16 km).

The energy for a thunderstorm is derived from the latent heat held by water vapor in the air. For that reason, thunderstorms are most common during summer months in the middle and higher latitudes, although they have been known to form in any season. Tropical regions have a high frequency of thunderstorms year round.

Thunderstorms are formed in three different ways: (1) thermals, (2) frontal activity, and (3) orographic circulation. *Thermally produced* thunderstorms (Figure 6–1) are the result of intense, localized convection. The rapid lifting of a column of warm, moist air (thermal) leads to condensation, which in turn releases latent heat of condensation. The latent heat adds additional heat to the air and accelerates the lifting of the air. Large cumulonimbus clouds may form from this activity and produce a localized storm. Usually this type of thunderstorm is very brief in any one location, lasting, on the average, less than 30 minutes. In that time, however, a very heavy downpour is characteristic, along with strong, gusty winds, thunder and lightning, and occasionally hail. Thermally produced thunderstorms are usually highly scattered across a region and are almost exclusively a feature of the summer season.

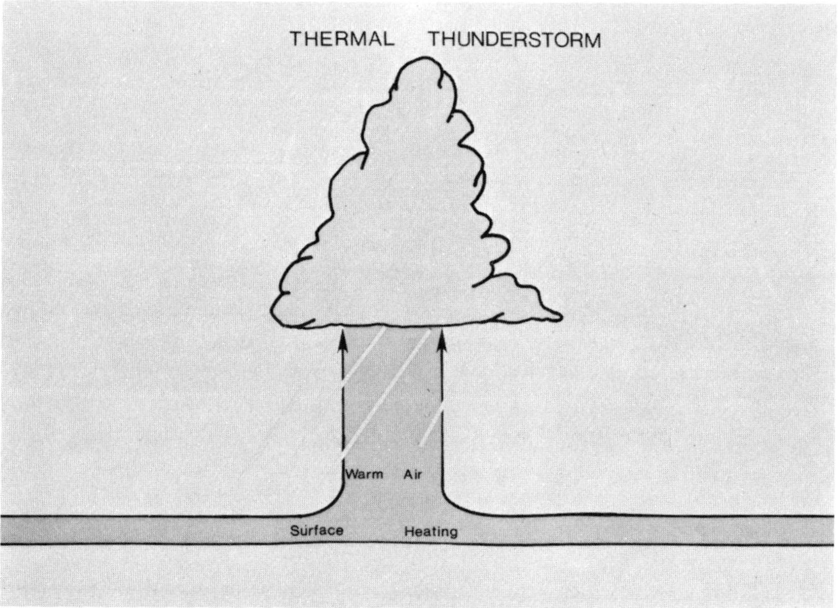

Figure 6–1

Thunderstorms produced by *frontal activity* form a more widespread network of storms (Figure 6–2). Frontal, or cyclonic, thunderstorms can be associated with warm or cold fronts in a middle-latitude cyclone, but most often they are a product of cold-front activity. As a cold front sweeps into a region occupied by warm, moist air, the warmer air is forced upward, triggering the rapid buildup of cumulus and cumulonimbus clouds along the entire length of the cold front. Unlike the thermal thunderstorm, which develops as an isolated individual storm, frontal thunderstorms often develop a nearly continuous line of storms along the leading edge of the cold front. Frontal thunderstorms rarely last more than one hour in any one locality, but

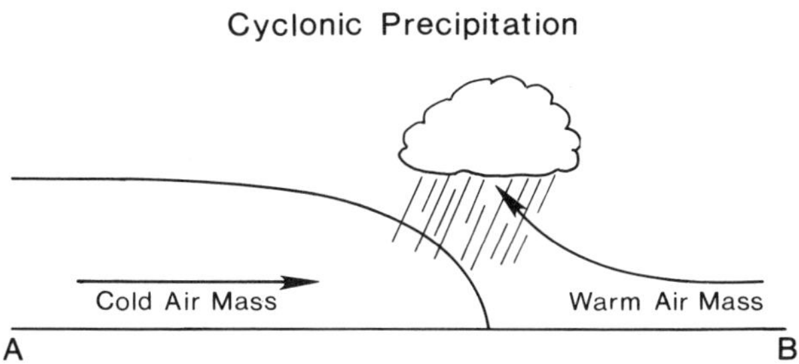

Figure 6–2

they can deposit abundant amounts of rainfall and are accompanied by strong winds, lightning and thunder, and perhaps hail.

The third method whereby thunderstorms are produced is *orographic lifting* of warm, moist air; this process was described in Module 4. What distinguishes an orographic thunderstorm from normal orographic rainfall is simply a matter of degree. The warm air must rise up windward slopes rapidly in order to trigger the rapid release of latent heat necessary to bolster the formation of cumulonimbus clouds. Orographic thunderstorms may be brief or widespread, depending on the nature of the air being lifted and the characteristics of the windward slope. As with all thunderstorms, large amounts of rainfall in brief, heavy downpours, along with gusty winds, thunder and lightning, and possibly hail, are generally characteristic.

Thunderstorms pass through a series of stages from their initial growth until they dissipate. These stages include (1) the cumulus stage, (2) the mature stage, and (3) the dissipating stage (Figure 6–3). In the initial stage of growth, called the *cumulus* stage, the storm is characterized by the formation of a rapid updraft of warm air which reaches the dew point, begins to condense, and releases latent heat of condensation. The latent-heat adds more energy to the updraft and a towering cumulonimbus cloud is formed.

The precipitation that forms in the cloud begins to fall within the cloud and produces a downdraft of air by friction. The *mature* stage has now been reached. The mature storm is characterized by a powerful updraft and a downdraft. When this downdraft reaches the earth's surface, it brings with it strong, gusty winds and abundant rainfall. This is the **squall** of the thunderstorm; it may join a line of other thunderstorm squalls to form a squall line. During the mature stage, cumulonimbus clouds may extend into the tropopause, where winds at that elevation blow the tops of the clouds downwind, giving the clouds an anvil-shaped top. The storm may remain in the mature stage if the updrafts continue to be fed by warm, moist air.

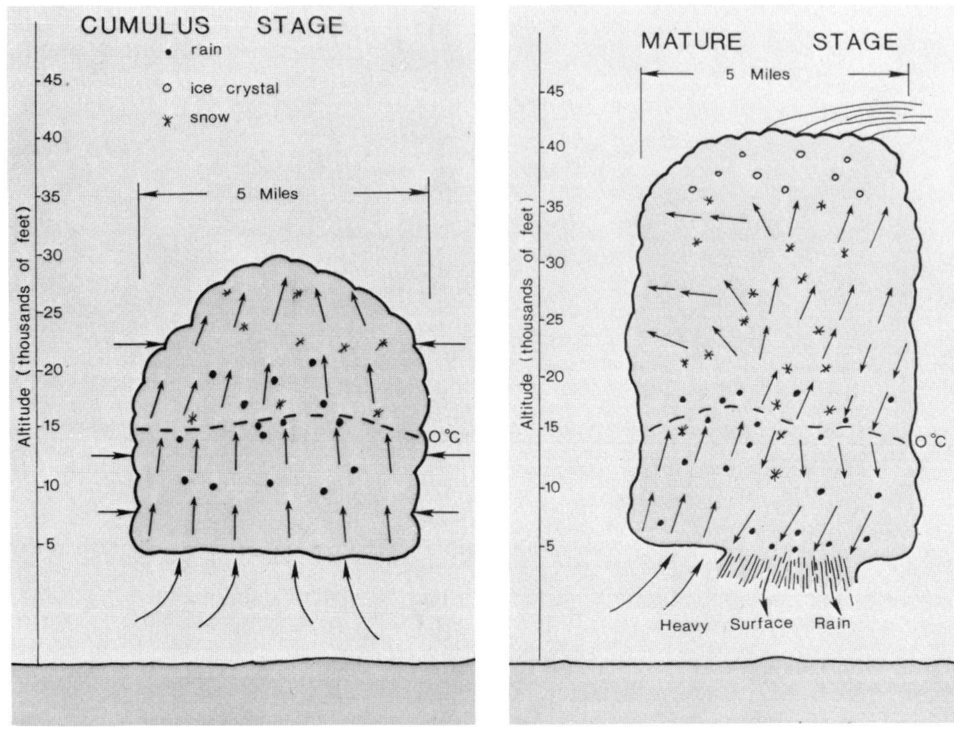

Figure 6–3

A thunderstorm enters the third, or *dissipation,* stage when the updrafts feeding the storm are cut off. The downdraft spreads across the entire base of the cloud and the storm quickly dissolves. Thunderstorms grow and dissolve in about one hour. As they move to new locations, however, fresh supplies of warm, moist air may supply the latent heat to renew the formation of updrafts and cause the thunderstorm activity to recharge itself as the cyclonic system moves along within the regional air circulation.

In North America, thunderstorms are most frequent in the summer months, east of the Rockies and south of the Canadian border. Figure 6–4A shows the average number of thunderstorms per year in the United States, and Figure 6–4B indicates the month with the highest number of thunderstorms. The increase in the number of thunderstorms along the east flank of the Rocky Mountains is due to the orographic lifting of invading maritime tropical (Gulf) air masses in the summer months.

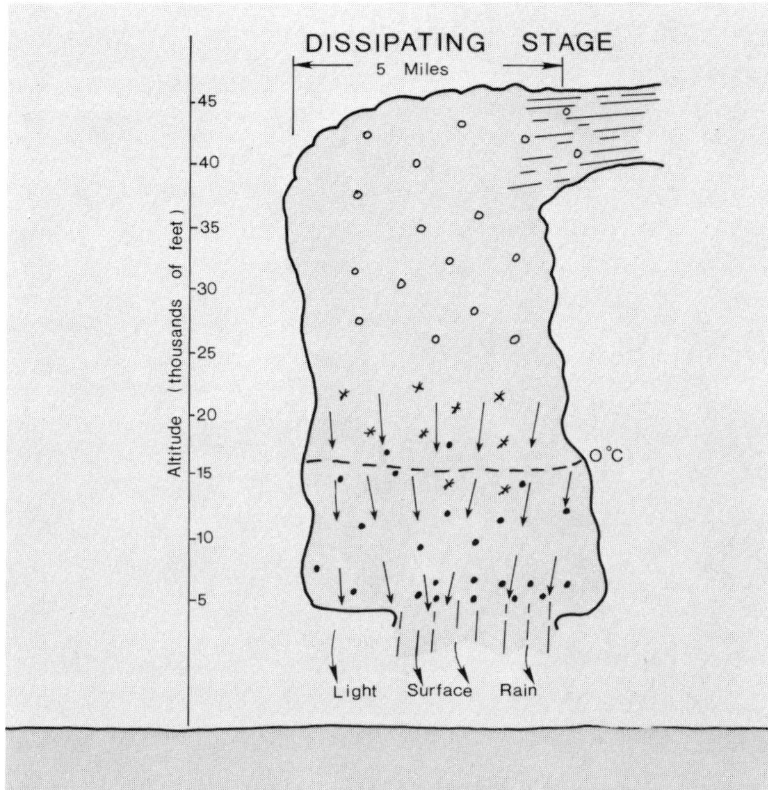

c.

Figure 6–3, Continued

Although scientists know how thunderstorms form, they are uncertain exactly how the thunder and lightning that always accompany thunderstorms are created. That is, they do not completely understand how massive concentrations of positive and negative electrical charges build up within clouds and between clouds and the ground. Some have proposed that the friction between ice particles or the formation of water droplets generates a negative charge in portions of a cloud; then a tremendous electrical charge is established within clouds and between clouds and the ground. These charges are estimated to reach several million volts before a **lightning** stroke is discharged. The air along the path of a lightning stroke is heated so rapidly that it literally explodes as it expands. The explosion of air sets up a shock wave in the air which is heard as **thunder.**

Thunder and lightning develop simultaneously but travel at different speeds toward an observer. One sees lightning at the speed of light, but sound

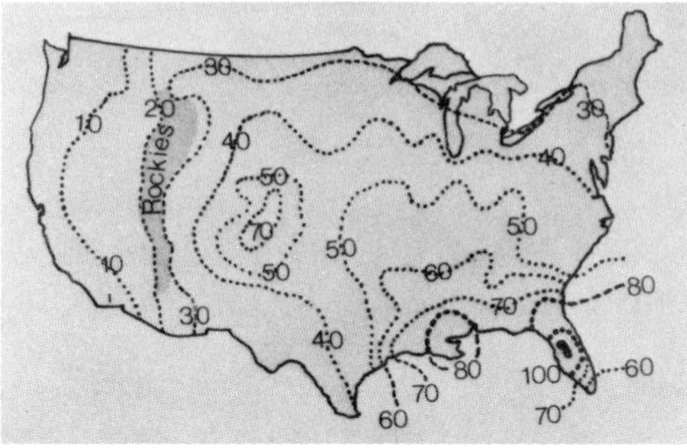

A.

B.

Figure 6–4

waves travel at a much slower rate (2000 feet/second). One method for determining the distance between you and a lightning stroke is to count the number of seconds that pass between the time you see the lightning stroke and hear the thunder, and then multiply the seconds by 2000 feet. On the average, lightning kills more people each year than any other single weather phenomenon: An estimated 300 to 500 people are killed annually by lightning strokes. Many of these deaths could probably have been prevented if proper precautions were taken.

Tornadoes

A **tornado** is a spinning region of intense low pressure; indeed, pressure in a tornado has been estimated to be 100 millibars (about 3 inches of mercury; see Module 3) over distances of only several hundred feet. Tornadoes develop in association with cold fronts and thunderstorms, and their energy is probably derived from latent heat in water vapor. Unfortunately, however, meteorologists are still uncertain as to the exact mechanism that causes these storms. Tornadoes appear to be directly related to the strong updrafts in a thunderstorm, but exactly why some updrafts produce a tornado while others do not—the key to understanding these storms—is not known with certainty. Keep in mind that although meteorologists are still uncertain exactly how tornadoes form, they do know the general conditions favorable for tornado development and therefore can usually issue tornado watches several hours in advance of an actual tornado formation.

Tornadoes are most frequent in the spring when air masses invading the middle latitudes possess great contrasts of temperature and moisture. Tornadoes are features almost exclusively of North America due to the ideal continental shape, which favors the continual clash of cold polar and warm subtropical air masses. Figure 6–5A shows the frequency of tornadoes in the United States. Oklahoma, Kansas, and Texas lead in total number of tornadoes per year, with the greatest concentration generally in the Great Plains and the least west of the Rockies and east of the Appalachians. Figure 6–5B shows the month when tornadoes are most frequent. Despite the number of tornadoes each year, the probability of a tornado is less than one per 100 square miles annually—even in Oklahoma.

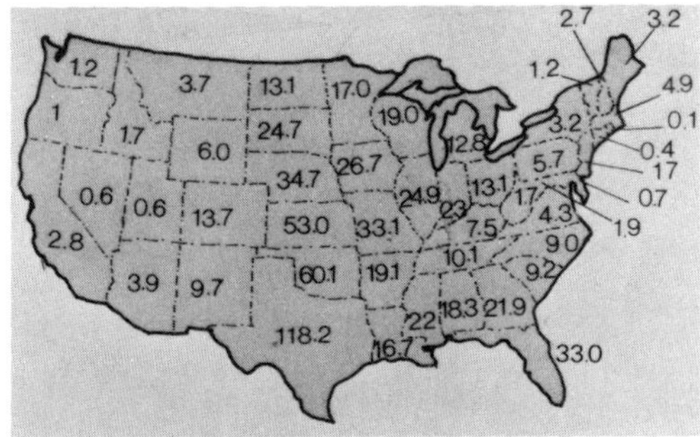

A.

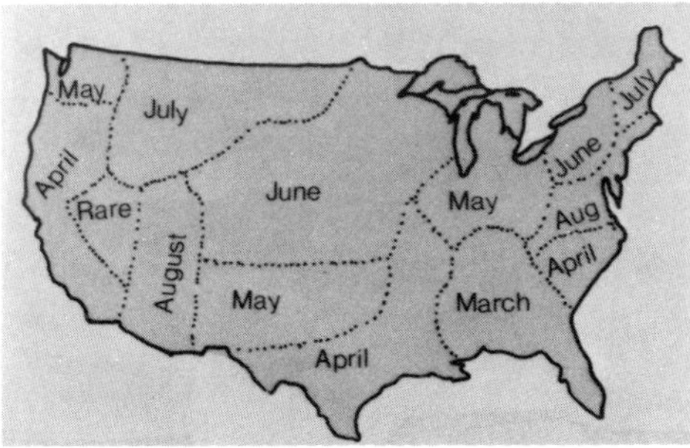

B.

Figure 6–5

Tornadoes develop along the leading edge of surface cold fronts directed by southwest winds. For that reason, more than two-thirds of the tornadoes recorded have moved in a southwest-to-northeast direction. Tornadoes have forward speeds that average near 30 miles per hour. The distance that a tornado may stay on the ground varies, but the average distance is 16 miles; some have been know to stay on the ground for distances more than 200 miles. The size of a tornado also varies greatly. "Mini-tornadoes" may be only several tens of feet across, whereas the "maxi-tornado" may be as much as 1½ miles across at its base. Fortunately, the occurrence of these monsters is quite rare.

Despite their relatively small size and infrequent appearances, tornadoes strike fear into the hearts of most people because of the destruction and deaths that they can cause. In a matter of seconds, homes may be completely demolished and people killed, despite sophisticated warning systems. However, the number of deaths attributed to tornadoes each year can be reduced if people follow the proper emergency procedures as outlined by the National Weather Service.

Strong winds exceeding several hundred miles per hour are often attributed as the cause for tornado damage. Recent tornado studies reveal, however, that most tornadoes do not have winds much in excess of 100 miles per hour. Much of the damage from tornadoes is caused more likely by the extreme pressure gradients that can occur between the inside and outside of a building as a tornado moves through an area. Pressures can be suddenly so much greater inside a building than outside that the walls of the building literally explode.

Whirlwinds of air that are like tornadoes but develop over water surfaces are called **waterspouts.** Waterspouts, however, are mild-mannered phenomena compared to tornadoes and represent a threat only to smaller boats.

Hurricanes

The third major severe storm is called the **hurricane** in the Western Hemisphere. In eastern Asia these storms are called *typhoons;* in India, *cyclones;* in Australia, *willy willys;* and in the China Sea, *baguios.* By whatever name, these storms are immensely powerful and are a threat to coastal inhabitants within a latitude belt 5° to 20° on either side of the equator.

Hurricanes are tropical low-pressure cells that form over tropical oceans and develop unusual strength. They derive their energy from the latent heat stored in the abundant water vapor available in the air over warm tropical oceans. Investigations reveal that hurricanes develop best when sea-surface temperatures exceed 80°F (27°C). Why some tropical cyclones develop into hurricanes and others do not is not yet completely understood, but once a hurricane is formed, it flows westward with the trade winds at speeds of 5 to 30 miles per hour. As hurricanes travel, the Coriolis force deflects them to the right in the Northern Hemisphere and to the left in the Southern Hemisphere.

Figure 6–6 shows the typical paths followed by hurricanes in the Northern Hemisphere. The average life span of a hurricane is eight to ten days, but one may last longer if its path coincides with a warm ocean current. Hurricanes in the Northern Hemisphere are most frequent in August, September, and October, when sea-surface temperatures are highest.

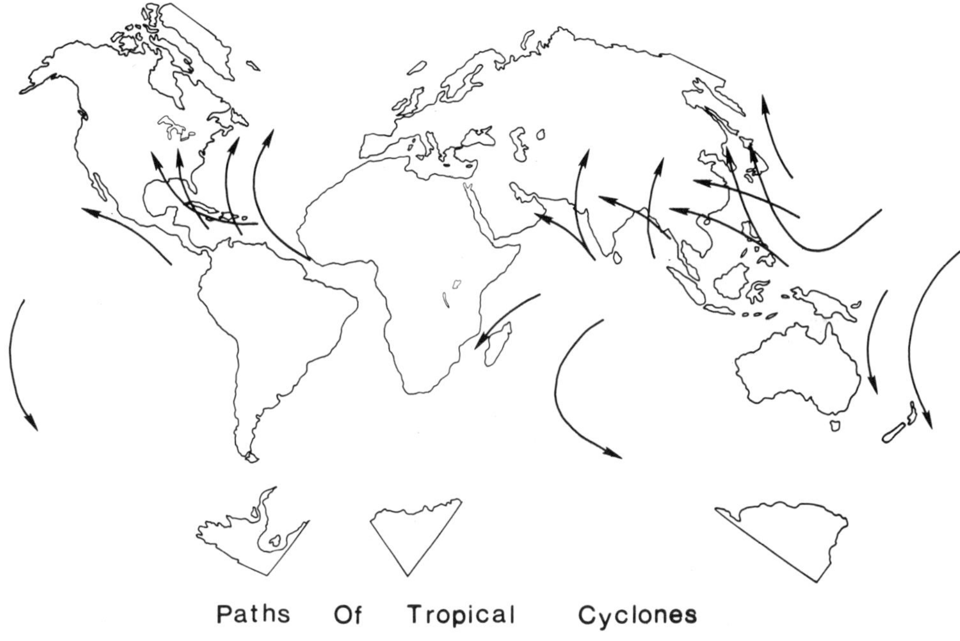

Figure 6–6

The National Weather Service classifies a cyclone as a *tropical storm* if it has wind speeds between 50 and 75 miles per hour. The storm becomes a *hurricane* when wind speeds reach 75 miles per hour. A hurricane—a spiraling mass of clouds and thunderstorms—may be 400 miles across and extend vertically to heights of 50,000 feet (Figure 6–7). The pressure gradients in a hurricane do not match those of a tornado, but pressure values may change from 50 to 60 millibars from the outer edge of the storm to the storm center. As long as the storm remains over a warm ocean, the warm, moist air continues to pump water vapor into the storm to provide the energy necessary to keep the storm active. Thunderstorms ring the center of the storm and may contain winds exceeding 100 miles per hour. At the very center of the storm is the area called the **eye.** This area is characterized by calm, almost cloud-free air which is slowly sinking from above. The spiraling air around the center of

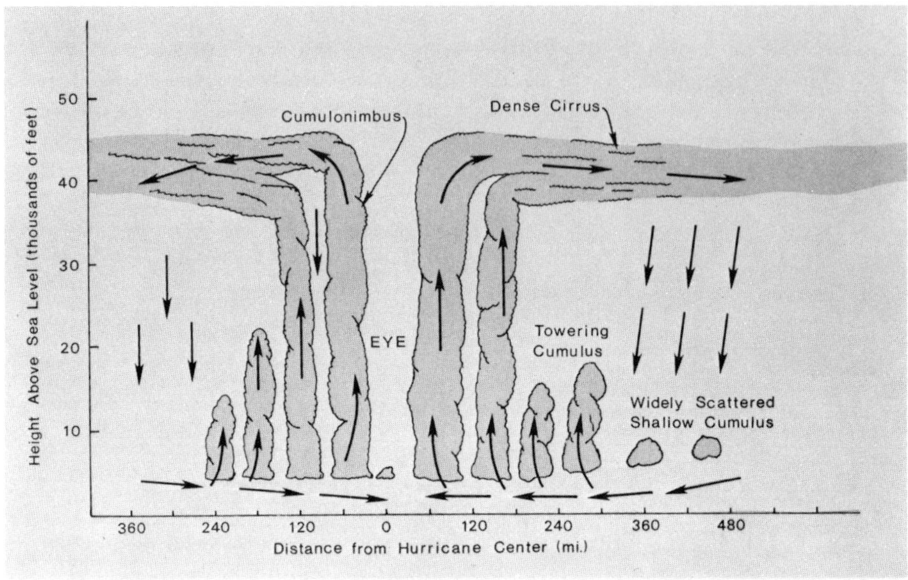

Figure 6–7

the storm is thrown outward from the center by centrifugal force, thus keeping the eye free of clouds.

When a hurricane strikes land, it brings with it torrential rains that may deposit 6 to 12 inches of water in only a few hours. High winds pound the coastlines and often demolish boats and harbor facilities. Most of the damage from hurricanes, however, is due to flooding, not high winds. When the storm is accompanied by high tides, the problems of flooding can be particularly acute.

Additional Material

Tornado Strikes Xenia, Ohio

On April 3 and 4, 1974, the Ohio Valley experienced the greatest one-day outbreak of tornadoes on record. That day, 148 tornadoes were spotted, more than twice the previous high of 69 on September 20, 1967. In one hour in north central Ohio, 16 tornadoes were sighted. Tornadoes covered an 11-state area, and their total path length was more than 2000 miles. (The average total path length in the entire United States in one year is 4000 miles.) Tornado paths were found in unexpected places such as in deep canyons and steep mountain slopes. The tornadoes developed out of three major squall lines that were moving from southwest to northeast. By 2:00 P.M., all three squall lines were growing at explosive rates. At 2:10 P.M., the first tornado was sighted in Illinois, and the outbreak was in progress.

Most of the tornadoes moved northeastward at a rate of 40 to 60 miles per hour, while the squall lines moved toward the southeast. This movement indicated that the tornadoes moved with individual thunderstorm cells, not with the squall lines. The storm system was a classic clash between warm, moist tropical Gulf air and cold polar air. The polar-front jet stream was sweeping from the southwest to northeast, intensifying the surface cyclone.

Fortunately, very few of these tornadoes struck major population centers. There was one terrible exception, however: Xenia, Ohio. At 4:15 P.M., a tornado struck the west edge of town and proceeded to rip a vicious path into the center of town. In the matter of just a few brief moments, 40% of Xenia was destroyed, including most of the business district and the local high school. The storm had traveled about 8 miles, totally destroyed 1297 buildings and badly damaged 3300 others, and killed 28 people.

Questions

1. Why are severe storms most common in the warmer months of the year in the middle and higher latitudes?

2. What are three methods by which thunderstorms can be initiated?

3. Why do thunderstorm clouds often have an anvil-shaped top?

4. How does a squall line develop?

5. How can radar help indicate the location of thunderstorms?

6. What prevents the formation of many thunderstorms west of the Rocky Mountains?

7. What causes a lightning stroke? How is lightning related to thunder?

8. Why are tornadoes a phenomenon primarily of the central section of the United States and not other regions in the world?

9. Why do most tornadoes form in the spring?

10. In addition to high wind velocities, what other factor is responsible for much tornado damage?

11. How do waterspouts compare with tornadoes?

12. What general ocean and atmospheric conditions are necessary for hurricanes to develop?

13. Why do hurricanes generally travel from east to west?

14. Why do hurricanes lose their energy when they move over land surfaces?

15. Some hurricanes move to latitudes of 45° or higher. What condition is necessary for the storm to move to such high latitudes?

Suggested Readings

ANTHES, R. et al. 1975. *The atmosphere*. Columbus, OH: Charles E. Merrill. Chapter 6.

BAKER, E. J., and PATTON, D. J. 1974. Attitudes toward hurricane hazard on the Gulf Coast. In *Natural Hazards,* by G. White. New York: Oxford University Press. Pages 30–36.

BATTAN, L. J. 1961. *Nature of violent storms*. Garden City, NY: Doubleday and Co.

———. 1964. *The thunderstorm*. New York: Signet Science Library.

———. 1974. *Weather*. Englewood Cliffs, NJ: Prentice-Hall. Chapter 6.

CHANGNON, S. A., JR., and SEMONIN, R. G. 1966. A great tornado disaster in retrospect. *Weatherwise* 19: 56–65.

DUNN, G. E., and MILLER, B. I. 1964. *Atlantic hurricanes*. Baton Rouge: Louisiana State University Press.

ISLAM, M. A. 1974. Tropical cyclones: coastal Bangladesh. In *Natural Hazards,* by G. White. New York: Oxford University Press. Pages 19–24.

LEHR, P. et. al. 1975. *Weather*. New York: Golden Press. Pages 96–111.

MALKUS, J. S. 1957. The origin of hurricanes. *Scientific American,* August 1957. Reprint 847. San Francisco: W. H. Freeman.

TEPLER, M. 1958. Tornadoes. *Scientific American,* May 1958. Reprint 848. San Francisco: W. H. Freeman.

TREWARTHA, G. T. 1968. *An introduction to climate*. 4th ed. New York: McGraw-Hill. Chapter 6.

Module
7

Hydrologic Cycle
and Local
Water Budget

INTRODUCTION

The planet *Earth* might better be termed the planet *Water*. Slightly more than 75% of Earth's* surface is covered by water, of which about 70% is in the oceans and an additional 5% is in lakes, swamps, and rivers. In addition, there are about 7 million cubic miles of potential surface water held in the form of glacial ice. Water is also held by soil, vegetation, and rocks, and in vapor form in the lower atmosphere.

---

*In this module, the astronomical convention of capitalizing *Earth* will be used.

107

Water is essential to all living things. We need water as much as we need air. We use water in virtually every facet of our everyday lives. Not only do we consume water directly to sustain our lives, but we also use water in a myriad of other ways: for agriculture, recreation, and navigation, and to dispose of waste and generate power.

Within the earth-atmosphere system, the total amount of water available in vapor, liquid, and solid form is, by and large, constant. This constant is maintained through a system called the *hydrologic cycle*. As you learned in Module 4, this cycle consists of the interaction of various processes: evaporation, condensation, precipitation, surface runoff, and absorption and flow of water in the soil subsurface. Basically, in the hydrologic cycle, water is evaporated from the oceans and is carried in the form of water vapor over a land surface where it condenses and falls to the ground as precipitation. A portion of the water falling on the land surface runs off the surface in rivers and streams; another portion is absorbed into the soil subsurface and slowly flows back to the sea; and yet another portion is held by lakes, glaciers, or vegetation, from which it is later evaporated and transported by winds moving back to the oceans. The result is an ongoing cycle of water from ocean to land and back to the oceans, which maintains a water balance on Earth.

The hydrologic cycle is a simplistic model of Earth's total water balance, but the balance of water in any *one* given area is more complex. In many cases, the supply of water in an area has changed dramatically with time. Several regions that once were covered with dense vegetation are now virtually deserts, while in other places just the reverse has happened.

The fluctuation in the distribution of water and the importance of water to humans has led to the development of many water specialists, such as meteorologists, oceanographers, glaciologists, limnologists, hydrologists, water engineers, water economists, and water conservationists. The supply of water, primarily fresh water, in a world of increasing population is a major cause of concern to many of these water experts. Not only is population increasing, but also the demand for water per person continues to increase, at least in the United States. In 1920, the average American used 120 gallons of water per day; by 1970, that figure had jumped to 170 gallons. Additional pressure on supplies is being applied in other sectors. Much more groundwater, for example, is being used to supply the increase in irrigation activity, while simultaneously, surface supplies of fresh water continue to be polluted by people and their activities.

Of most importance to humans are the local distribution and supply of water. Will there be an ample supply of water to raise my crops? Where will I get my drinking water? Is the local water supply fit for human consumption? These questions, and many more, are asked by people every day around the world. The need for fresh water continues to escalate. It is essential that we conserve and maintain the quantity and quality of fresh water.

OBJECTIVES

By the end of this module, you should be able to do the following:

1. Describe the various phases that compose the hydrologic cycle.
2. Explain the factors that are responsible for the variations in soil moisture supplies.
3. Describe and explain the physical features that determine the variations in depth of groundwater in the subsurface and the quantity of groundwater available.
4. Describe the relationship between groundwater and surface water supplies.
5. Describe the general importance of groundwater to humans.
6. Describe and explain the various features of the local water budget.
7. Describe how the above features vary with the seasons.

KEY TERMS

hydrology	peizometric surface
zone of aeration	artesian pressure
zone of saturation	recharge area
water table	influent stream
aquifer	effluent stream
unconfined aquifer	spring (seep)
confined aquifer	geyser
base flow	field capacity

Now you are ready to begin the audiovisual portion of this module. Select the MEDIAPAK 7 component(s) and proceed. Following is a topical outline of the audiovisual sequence. You will find this outline helpful for reference and review. After completing MEDIAPAK 7, return to this book to perform the exercises.

OUTLINE

Hydrologic cycle

Water held beneath the surface
Soil moisture
    Factors that cause variations
Groundwater
    Factors that cause variations
    Relationship to surface water
    Human use

Local water budget
Seasonal variations

Exercises

1. Which of the following soils will probably have the greater infiltration rate?
   A. Sand and gravel
   B. Clay

2. There is more water on Earth in the form of groundwater than there is available as fresh water in rivers, lakes, and streams.
   A. True
   B. False

3. The zone of saturation includes the layer of soil moisture.
   A. True
   B. False

4. The depth of the water table may depend on
   A. the amount of moisture available from the atmosphere.
   B. the presence or absence of unconsolidated materials.
   C. the depth of the porous material that holds moisture.
   D. all of the above.
   E. only A and C above.

5. The rate of flow of water in a well may be totally unrelated to the moisture that falls to the ground at the well site.
   A. True
   B. False

6. If water in a well is higher than the water table, this water level is called the
   A. artesian level.
   B. recharge surface.
   C. aquifer.
   D. piezometric surface.

7. Letter *A* of the following diagram represents a(n)
   A. influent stream.
   B. effluent stream.
   C. spring.
   D. artesian well.

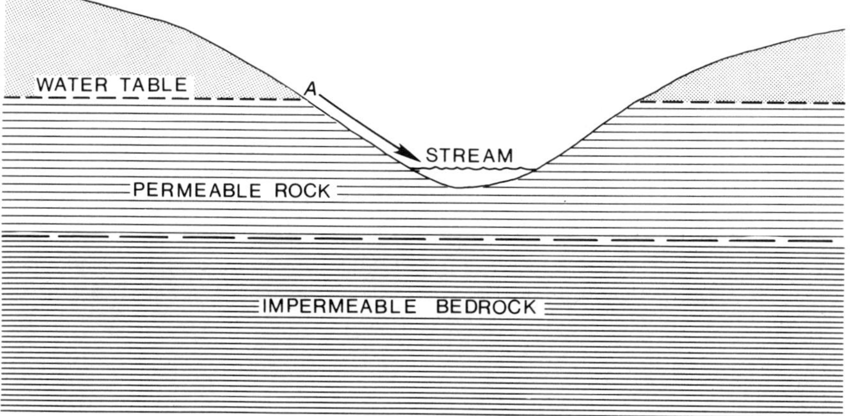

8. Influent streams are most common in
   A. dryland areas.
   B. humid areas.

9. What percentage of U.S. cities derive their water supplies from wells?
   A. 10%
   B. 25%
   C. 50%
   D. 75%
   E. 90%

10. For a local water budget in the middle latitudes, the supply of moisture fluctuates with the seasons. Which of the following statements regarding such a local water budget is/are true?
    A. The winter months are a time of increased storage of moisture in the form of snow.
    B. Soil moisture supplies are often at field capacity in the spring.
    C. Moisture supplies are rapidly depleted in the fall due to reduced rainfall but continued warm temperatures.
    D. All of the above.
    E. Only A and B above.

Summary

The total amount of water within the earth-atmosphere system is relatively constant due to the hydrologic cycle. This cycle graphically displays the flow of water in its various states as it is transported from place to place. One part of the hydrologic cycle is the portion of water that falls on land, the study of which is called **hydrology.**

When rain falls to the earth's surface, some is immediately evaporated back into the air, some runs off the land in streams and rivers, and some is absorbed into the subsurface as groundwater storage. Much of the absorbed moisture is retained by soil. Soil moisture supplies vary tremendously from place to place due to variations in soil type, climate, and vegetation. Soil moisture is essential for plant growth since most plants draw moisture that they need from the soil. Soil moisture supplies are diminished by direct evaporation from the soil surface and by transpiration from plant leaves. A large tree on a hot summer day, for example, may transpire several hundred gallons of water.

Water that infiltrates below a depth where it can be utilized by plants is called *groundwater*. Groundwater supplies are estimated to equal 15% of the amount of fresh water stored in glacial ice, or 30 times more water than is in all of the rivers, lakes, and streams in the world.

The subsurface is divided into two zones of moisture. The topmost layer is the zone of soil moisture or **zone of aeration;** the second layer is the **zone of saturation,** which includes the water table. The **water table** marks the top surface of the zone of saturation. The depth of the water table varies with location due to the amount of available moisture, the presence or absence of unconsolidated materials, and the depth of the porous material that holds the moisture. Some layers of rock and gravel in the zone of saturation are especially well suited to holding water; these layers are called **aquifers.**

The majority of aquifers that supply wells in the United States are composed of sand and gravel. Receding glaciers or ancient deposits of sediments are generally the source for sand and gravel aquifers. Other major aquifers are located in beds of sandstone or limestone. Aquifers vary not only in composition but also in size. Sand and gravel aquifers may be 200 feet (650 m) thick, while sandstone or limestone aquifers may reach thicknesses of 1000 feet (3280 m) or more. Large aquifers may extend horizontally to cover several thousand square miles.

Aquifers are either confined or unconfined (see Figure 7–1). If the rock strata above an aquifer allow water to pass through them, the aquifer is **unconfined.** On the other hand, if the rock layers above an aquifer are impermeable to water, the aquifer is **confined.** In a confined aquifer, a horizontal water table at a given depth does not exist because the confining rock layers are usually sloped rather than horizontal. As a result, the water pressure is much higher in the low end of the aquifer. If a well were sunk into the low end of a confined aquifer, the high pressure would force water to rise much higher in the well than the general water table level. The level to which the water pressure rises is called the **piezometric surface.** The water pressure in confined aquifers is called **artesian pressure.** Water may rise to the surface in a well due to artesian pressure, and the well is called an *artesian well.*

The water in an aquifer is derived generally from a large region called the **recharge area.** Water may travel several hundred miles before reaching the location at which it is withdrawn. Precipitation amounts at the site of a well, consequently, do not necessarily determine the amount of water that will be available in the well. Eastern Nebraska, for example, is a semiarid region, but there are many artesian wells in that part of the state thanks to the Rocky Mountains to the west which serve as the major recharge source.

The movement of water in an aquifer varies from a few feet to several miles a day. As water infiltrates an aquifer, much of the mineral content is withdrawn by the rock material. Therefore, most aquifers possess fresh water. In certain regions, however, the presence of particular minerals (iron, for example) or the proximity of ocean water can contaminate the aquifer. The

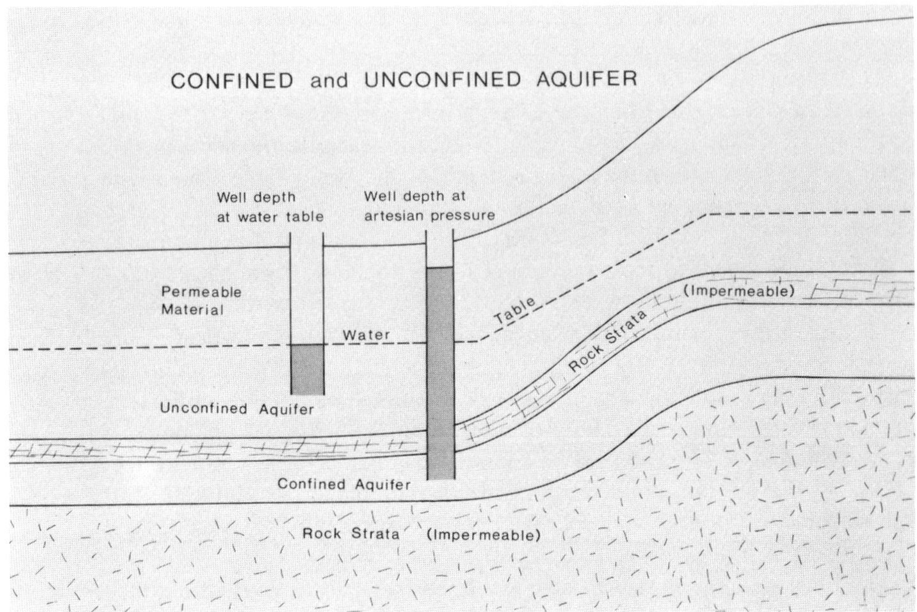

Figure 7–1

water in a large aquifer with a low flow rate may be stored for more than 100 years.

Where the slope of the aquifer and the slope of the surface intersect (Figure 7–2), the water in the aquifer may flow into a surface stream. This type of stream is termed an **influent stream.** Influent streams usually continue to flow even under extreme drought conditions, when other streams may dry up. The amount of water in a stream that is supplied by groundwater is called the **base flow** in that stream. Another groundwater-related stream is the **effluent stream,** which supplies water to aquifers. Generally, effluent streams are features of humid climates, whereas influent streams are more common in arid climates. Some streams may exhibit both influent and effluent characteristics at different seasons of the year.

Springs, or **seeps,** are groundwater features that are the result of a discharge of groundwater onto the surface at a level above a surface stream. The rate of water flowing from different springs varies. For example, some large springs may supply enough water to satisfy the needs of a small community, whereas smaller springs may flow only during the wetter seasons of the year.

The water temperature of a spring may be near freezing or boiling, depending on the location of the spring's source. Springs that are drawn from deep aquifers that are in close proximity to molten rock may reach boiling

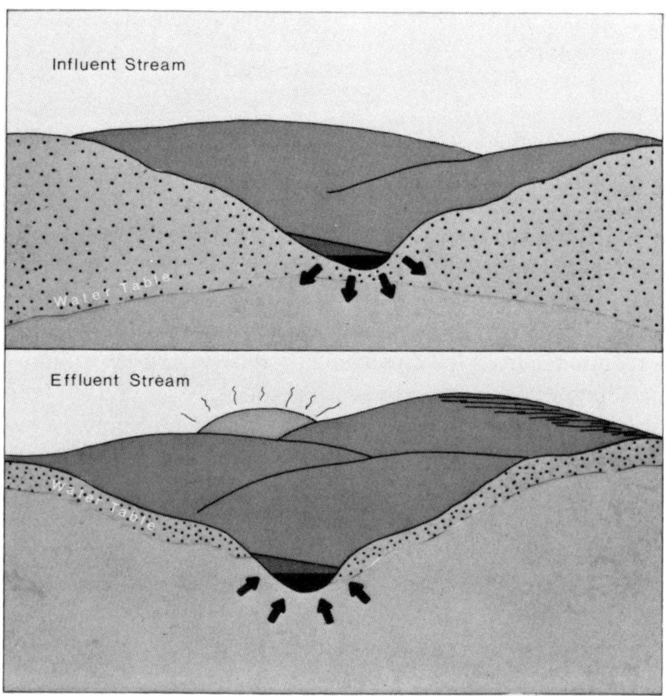

Figure 7–2

temperature. These are thermal springs. The hot water in a thermal spring may emerge at the surface under great steam pressure and burst into the air and form a **geyser.** Old Faithful, in Yellowstone National Park, is probably the best-known geyser in the world. Old Faithful blows a spectacular fountain of steaming and boiling water more than 100 feet (328 m) into the air about once every 60 minutes. Not only are geysers often spectacular features, but also they may serve an important economic function, powering generators for electricity.

Cold springs tend to have a source much closer to the surface. Mineral springs, which may be hot or cold, contain a high concentration of minerals which many people believe have a certain health benefit.

Goundwater is a valuable resource. By the late 1960s, it was estimated that more than 22% of the water used in the United States came from groundwater. It supplements local water supplies that fill residential, industrial, and agricultural needs. Irrigation is the largest single user of groundwater and is particularly important in the 17 western states. In fact, the increase in irrigation has caused some concern as to the possible depletion of the supply of groundwater. Indeed, even though recent estimates suggest that a 75-year supply is available in the United States, certain areas have already experienced periodic shortages.

Local Water Budget

The hydrologic cycle demonstrates that moisture tends to maintain a worldwide balance between sea, land, and air. On a local scale, however, there may be considerable fluctuation between moisture supply and moisture demand over a period of time. There may be periods of drought or times of flooding. A knowledge of the local water budget is used to forecast the potential requirements of local residents, industry, and agriculture.

There are four major components of the local water budget: precipitation, soil moisture, evapotranspiration, and runoff. In some regions, snow accumulation and melting are also considered. The local water budget is in balance when the moisture entering the area is equal to the amount of moisture held temporarily in the area plus the amount leaving the area. Precipitation and melting snow represent the total supply to the area; soil moisture represents the amount of water temporarily held in the area; and evapotranspiration, groundwater, and runoff represent the total outflow:

Local Water Balance

Precipitation + melting snow
          = soil moisture + evapotranspiration + groundwater + runoff

---

If the total moisture outflow exceeds the total moisture inflow, the area experiences a net moisture deficit. Inflow cannot exceed outflow because the excess moisture goes either to groundwater supply or to increased runoff.

The seasonal variation in the moisture balance is reflected in the fluctuation of soil moisture, the primary storage mechanism in the water budget. Soil has a maximum moisture-storage capacity called **field capacity.** When plants are transpiring at maximum rates and the soil moisture is at field capacity, moisture applied to the land is excess and will either run off the surface or recharge the groundwater supply. When the amount of precipitation is less than the potential evapotranspiration, moisture will be withdrawn from the soil and the total supply will be less than field capacity. The soil moisture balance is, therefore, a reflection of the seasonal changes in the local water balance.

Let's briefly analyze one local water-balance situation—in the midwestern states of the United States (Figure 7–3). In the middle latitudes, the moisture fluctuations during the winter are primarily from melting snow. The soil moisture remains relatively constant because the ground is frozen and the

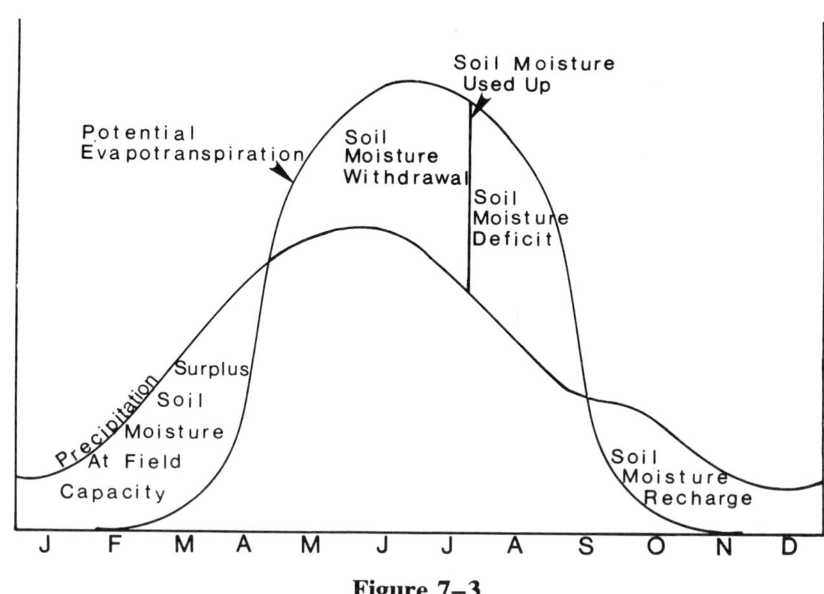

Figure 7–3

soil moisture is generally near field capacity. Overall, the winter months are a period of increased water storage in the form of snow.

In the spring, water from melting snow and spring rains raises soil moisture supplies to field capacity. Relatively cool temperatures at this time of year, along with the absence of a full vegetation cover, limit evaportranspiration. Thus, spring is a period of moisture excess; runoff amounts are large and groundwater supplies are recharged.

Summer marks a period of great moisture fluctuations. Evapotranspiration amounts are large due to high temperatures and full vegetation growth. Without regular additions of precipitation, the soil moisture supply will be reduced and may be almost exhausted if too many weeks pass without adequate rainfall. On the whole, soil moisture supplies are reduced in the summer, as are groundwater supplies.

Fall is a time of soil moisture recharge. Precipitation amounts are usually sufficiently large to exceed the potential evapotranspiration. Consequently, soil moisture supplies are gradually replenished and groundwater is slowly recharged. Thus, in some sections of the Midwest, there are periods of moisture excess and moisture deficiency.

Around the world, the pattern of moisture supply varies with place and time. Classifications of world climates have been developed that recognize the imbalances between regions where the annual precipitation amount exceeds the potential evaporation and vice versa.

Questions

1. What factors determine the depth of the water table in an area?

2. What is the difference between a confined and an unconfined aquifer?

3. Why does the concept of a water table not really apply in a confined aquifer? What is the relationship between the water table and the piezometric surface?

4. How does an artesian well differ from a regular well?

5. How does a spring differ from an effluent stream?

6. Do geysers have a practical value? Explain.

7. When and why do surpluses and deficits occur in the soil moisture supply in the middle latitudes?

8. When does soil moisture recharge take place? Why?

9. What reasons can you list for a drop in the water table?

Suggested Readings

LEET, L. D., and JUDSON, S. 1958. *Physical geology.* 2d ed. Englewood Cliffs, NJ: Prentice-Hall. Chapter 11.

McINTYRE, M. P. 1966. *Physical geography.* New York: Ronald Press. Chapter 8.

OLIVER, J. E. 1977. *Perspectives on applied physical geography.* North Scituate, MA: Duxbury Press. Chapters 11 and 12.

OLSON, R. E. 1970. *A geography of water.* Dubuque, IA: William C. Brown. Chapter 7.

SAYRE, A. N. 1950. Ground water. *Scientific American,* November 1950. Reprint 818. San Francisco: W. H. Freeman.

Module
8

World
Climatic
Regions

INTRODUCTION

In the past several modules, we examined in some detail each of the elements that determine weather in the atmosphere. We analyzed the composition and variations of each of these climatic elements *individually*. Now we will consider the climatic elements in *combination* with one another. In particular, we will examine how these elements interact from place to place around the world. Our frame of reference will be the region—a unit of area in which associated features relate to one another in about the same manner throughout,

119

giving the region an internal unity that distinguishes it from other regions. A *climatic region* is an area in which the relationship between the various climatic elements is about the same throughout the entire region. For example, within a given range, all places with the same climatic classification have about (1) the same amount of precipitation with a similar pattern of distribution; (2) the same general temperature range with similar daily, monthly, and yearly variations of temperature; and (3) the same pressure pattern and wind systems. When analyzing world climates, we must answer the following two basic questions:

1. What are the various associations between climatic elements
    that form unified climatic regions?
2. How do climatic regions vary from place to place?

Each region has its own characteristic climate. Indeed, no two places on the earth have exactly the same climate. The climate on top of a high hill is slightly different than it is at the bottom of the hill. Similarly, the climate varies from the top of a tree to the base of the tree. In fact, it would be possible to describe the climate of each square foot on earth; however, we would have such a maze of different climates that any discussion of climates on a large scale would be meaningless. Instead, geographers have noted that within certain regions, the climate is roughly similar throughout, with only very minor changes in one or several climatic elements. To analyze worldwide changes in climate, geographers must temporarily disregard the minor fluctuations of climate and search for broadly similar climatic characteristics. By establishing broad limits for a given climate, they have been able to classify groups of similar climates on a worldwide scale.

The ancient Greeks recognized three broad types of climates, which they called the *torrid zone,* the *temperate zone,* and the *frigid zone.* This simple division of world climate was based on the relationship between temperature and latitude, and it is still a correct view of temperature, as far as it goes. But we know now that more details are needed for even the simplest usable framework of world climates. Most important, the moisture supply must be considered. The Greeks perceived the warmest and coldest regions correctly, but they failed to distinguish the wet climates from the dry ones.

From the list of climatic elements, many climatic classifications have been developed, some of which are better for specific purposes than others. Perhaps the most common classification used by geographers today is the one developed in the early 1900s by Wladimer Köppen, a European biologist and climatologist. Recognizing that natural vegetation reflects climatic conditions, he established classes of climate whose limits correspond to the temperature and moisture requirements of various plant types. Köppen's original classification is highly detailed and includes many subclasses, and so several revised and simplified classifications of climates have been developed from his original model. In fact, virtually all of the more recent classification systems are modifications of Köppen's system. Our discussion of world climates will also be a simplified modification of Köppen's.

OBJECTIVES

By the end of this module, you should be able to do the following:

1. Describe the distinguishing climatic characteristics of each of the 13 general climatic regions of Köppen's classification.
2. Explain the factor(s) responsible for controlling each of the 13 climate regions.
3. Recognize on a world map the general locations of the 13 climate regions.

KEY TERMS

climograph   mirage

ITCZ   climatic regions

Now you are ready to begin the audiovisual portion of this module. Select the **MEDIAPAK 8** *component(s) and proceed. Following is a topical outline of the audiovisual sequence. You will find this outline helpful for reference and review. After completing* **MEDIAPAK 8**, *return to this book to perform the exercises.*

OUTLINE

Climographs

World climatic regions
Humid tropical climates (A)
    Tropical rainforest (Af)
    Tropical monsoon (Am)
    Tropical savanna (Aw)
Dry climates (B)
    Steppe (BW)
    Desert (BS)
Humid mesothermal climates (C)
    Mediterranean (Csa, Csb)
    Humid subtropical (Caf)
    Marine west coast (Cbf)
Humid microthermal climates (D)
    Humid continental (Daf, Dbf)
    Subarctic (Dcf, Ddf)
Polar climates (E)
    Tundra (ET)
    Icecap (EF)
Highland (H)

Exercises

1. Which of the following descriptions does not characterize the tropical rainforest climate?
    A. Dominated by equatorial low-pressure belt all year.
    B. Smallest annual range of temperature of any climate.
    C. Precipitation maximum associated with position of subtropical high.
    D. Daily range of temperature exceeds annual range.
    E. No dry seasons during year.

2. The dry season in the tropical monsoon climate is associated with
    A. dry offshore winds triggered by a seasonal change in pressure.
    B. cold offshore ocean currents.
    C. sinking air from aloft in the subtropical high.
    D. continental location.
    E. leeward position along large mountain range.

3. In the savanna climate, the six-month dry season can be attributed to the dominating presence of the ITCZ.
    A. True
    B. False

4. Which of the following descriptions is not a characteristic of the dry climates?
    A. Leeward position along mountain range.
    B. Continental location.
    C. Daily range of temperature highest in desert climates.
    D. Calm desert days and windy desert nights.
    E. Precipitation totals generally less than 25 inches annually.

5. Which climatic region is controlled by warm offshore ocean currents and westerly winds?
    A. Tropical rainforest
    B. Humid continental
    C. Subarctic
    D. Desert
    E. Marine west coast

6. The greatest annual range of temperature is a characteristic of the
    _____________________________ climatic region.
    A. tundra
    B. tropical savanna
    C. subarctic
    D. desert
    E. tropical monsoon

7. Which climatic region lies along the east coasts of continents between major oceanic subtropical high-pressure cells?
    A. Humid continental
    B. Humid subtropical
    C. Tropical monsoon
    D. Steppe
    E. Tropical savanna

8. Which climatic region has only two to four months with average monthly temperatures above freezing, but all months with average monthly temperatures below 50°F?
   A. Icecap
   B. Subarctic
   C. Tundra
   D. Steppe
   E. Desert

9. Which climatic region often records the highest annual precipitation totals but also has a two-to-three–month dry period?
   A. Tropical savanna
   B. Steppe
   C. Tropical rainforest
   D. Tropical monsoon
   E. Humid subtropical

10. Which climatic region is controlled by subtropical high pressure during summer, and westerly winds and cyclonic precipitation in winter?
    A. Mediterranean
    B. Humid subtropical
    C. Tropical monsoon
    D. Steppe
    E. Tropical savanna

11. The Amazon basin and the Zaire basin (Congo) are the two largest continuous regions of _______________________ climate.
    A. tropical monsoon
    B. humid subtropical
    C. tropical rainforest
    D. tropical savanna
    E. humid continental

12. The fluctuation of the ITCZ and the subtropical highs each six months is the major control of the _______________________ climatic region.
    A. tropical monsoon
    B. steppe
    C. humid subtropical
    D. Mediterranean
    E. tropical savanna

Summary

Although no two places on the earth have exactly the same climate, geographers have noted that within certain regions, the climate is roughly the same throughout, if minor variations of one element or another are discounted. By establishing broad limits for climates, geographers have been able to divide the world into a pattern of various climatic regions; and by so doing, they have facilitated the analysis of large-scale variations in climate. Despite this convenience, one must not lose sight of the fact that small-scale variations within each subclimate are of extreme importance when analyzing local areas.

Many of the various climatic classifications in use today were derived from the classification done by Wladimer Köppen in the early 1900s. The classification used in this module is a modification of Köppen's original classification. It divides the world into major climatic regions, which are divided into 13 subclimates, each of which is labeled with a descriptive title and a letter symbol. These 13 subclimates are listed in Table 8–1. Each of them represents a special region on the earth where a unique atmospheric environment is created by the interaction of the climatic elements, a combination that is not duplicated within any other climatic region. You should be able to describe the interaction of the climatic elements within each region and explain the factors responsible for such patterns of interaction.

Table 8–1

Humid Tropical Climates (A)
 1. Tropical rainforest (Af)
 2. Tropical monsoon (Am)
 3. Tropical savanna (Aw)

Dry Climates (B)
 4. Desert (BW)
 5. Steppe (BS)

Humid Mesothermal Climates (C)
 6. Mediterranean (Csa, Csb)
 7. Humid subtropical (Caf)
 8. Marine west coast (Cbf)

Humid Microthermal Climates (D)
 9. Humid continental (Daf, Dbf)
 10. Subarctic (Dcf, Ddf)

Polar Climates (E)
 11. Tundra (ET)
 12. Icecap (EF)

13. Highland Climate (H)

Temperature and precipitation are the two key elements that define the limits of each climatic subclass. These two elements can be graphically displayed on a **climograph** to illustrate the annual variation of each element within a subclimate. Figure 8–1 is a climograph of Madison, Wisconsin, with mean monthly temperatures plotted on the left scale and mean monthly precipitation plotted on the right scale. A typical climograph for each of the subclimates is given in the Additional Material section of this module.

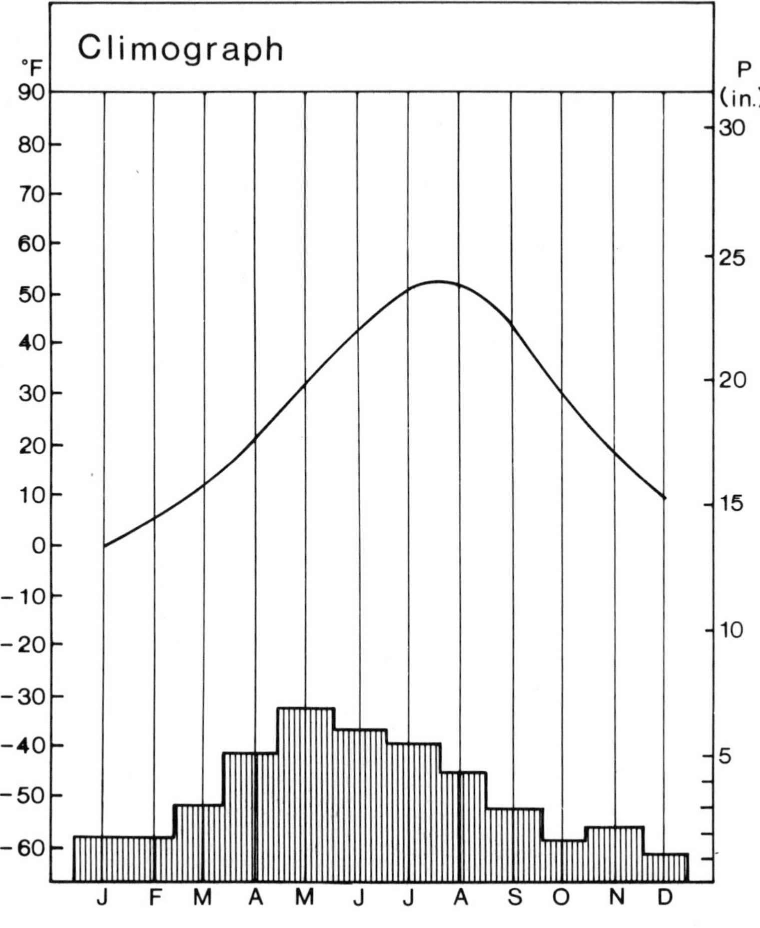

Figure 8–1

Humid Tropical Climates (A)

The first humid tropical climate is the *tropical rainforest* (Af). It is characterized by a lack of seasonal change, and each day is much a duplicate of every other day. Temperatures are high, humidity is high, and precipitation is abundant. On the world climate map in Figure 8–2, the locations of the tropical rainforest climate are shown. The largest regions are the areas surrounding the Amazon River basin and the Zaire River basin (Congo River). Monthly average temperatures exceed 65°F (18°C) throughout the year, and the annual range is the smallest of any subclimate (less than 10°F). Daily maximum temperatures are generally in the 80s, and evenings often cool into the 60s. The cool evenings have been labeled the *winter* of the tropics.

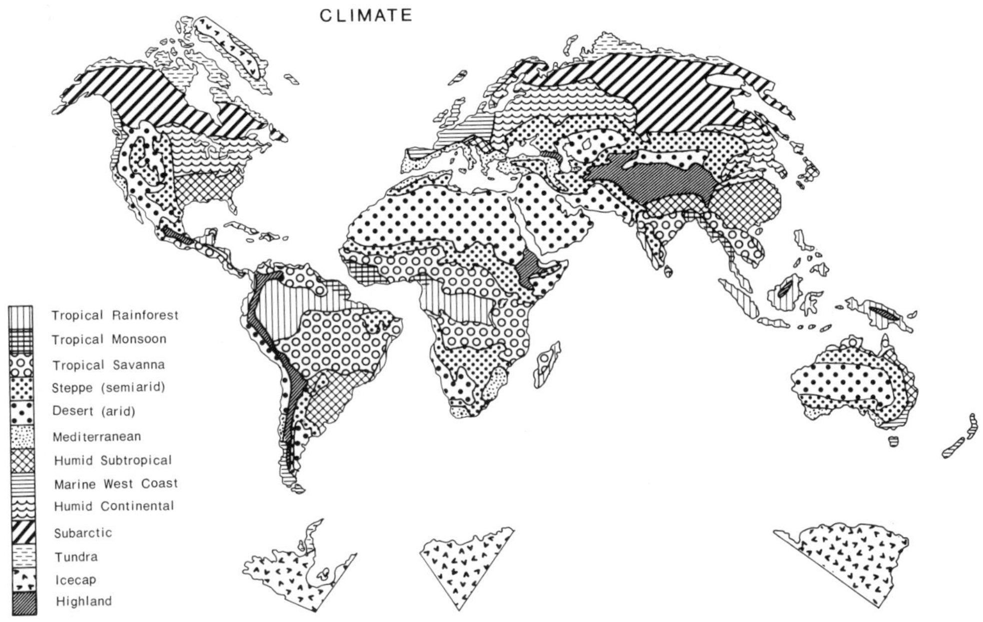

Figure 8–2

Precipitation in the tropical rainforest climate is abundant and uniformly distributed during the year. Most locations receive more than 60 inches (152 cm) per year. No season can truly be called *dry,* although in some places there may be more rain in some months than others.

The tropical rainforest climate is located generally within 15° of the equator. In this belt of equatorial low pressure, converging air circulation near the surface is common. Embedded in the approximate center of the equatorial low is a zone where convergence is most active. This zone, called the **Intertropical Convergence Zone (ITCZ),** triggers the uplift of warm, moist air at the surface, which results in large amounts of rainfall. The ITCZ shifts its position during the year in direct response to the movement of the sun's vertical ray. Since the ITCZ is a major cause of heavy precipitation, the latitudinal position that it occupies at any one time largely determines the region of maximum precipitation. Most of the rainfall triggered in the ITCZ and throughout the tropical rainforest region is in the form of brief, localized thundershowers. The general convergence at the surface usually results in a considerable cloud buildup, even though no rain may be produced. Clouds help reduce the amount of insolation, which in turn keeps temperatures from reaching daily maximums in the 90s or above, temperatures that are more common slightly farther poleward.

The second tropical climate, the *tropical monsoon* climate (Am), differs from the tropical rainforest climate in that it has a characteristic two- or three-month annual dry season. Tropical monsoon regions are located between 0° and 20° latitude and generally along coastal areas. Seasonal reversals in wind direction due to pressure changes are responsible for the distinct wet and dry periods in monsoon regions. Low-pressure cells are most common over land surfaces in summer and result in onshore ocean winds that transport large quantities of evaporated moisture to the land. Onshore winds are the source of the abundant precipitation that creates a wet season. In the winter, high pressure over the land causes winds to blow out to sea. Thus, the dry air drawn from the interior regions of large landmasses limits precipitation and results in several months without rainfall.

Average annual rainfall in monsoon areas matches or exceeds that of the tropical rainforest regions. In fact, monsoon regions often receive more total rainfall than any other climate. Annual amounts may exceed 200 inches (500 cm), although the average is closer to 100 inches (250 cm). During the peak of the wet season, one month may receive a total of more than 30 inches (76 cm), an amount equal to the yearly total for many middle-latitude locations.

Temperatures in the tropical monsoon regions are high year round, with monthly averages in excess of 65°F (18°C). The annual range is slightly greater (10° to 15°F) in monsoon regions than in tropical rainforest regions because clear skies during the dry months allow average daily temperatures to exceed 80°F.

The third humid tropical climate is the *tropical savanna* climate (Aw). Tropical savanna regions form the boundary between the humid tropical regions and the tropical semiarid and arid regions. As shown in Figure 8–2, the savanna regions lie on the interior and poleward margins of the humid tropical climates (Af, Am). The savanna climate is located between latitudes of 5° and 20°.

The major feature of the savanna regions is that they experience about six wet months and six dry months each year. The control of this alternating pattern is alternating pressure or, more precisely, the fluctuation of the ITCZ (equatorial low) and the subtropical highs during the year. As the pressure belts shift position during the year in response to the movement of the sun's vertical ray, a region may come under the influence of the equatorial low-pressure belt and the ITCZ. At that time, rainfall will be abundant in that region and a wet season will result. As the equatorial low and the ITCZ shift away from a region, however, they may be replaced by subtropical high pressure. Under high-pressure domination, the precipitation mechanism is largely destroyed, thus initiating a dry season. The total annual rainfall averages 40 to 60 inches (100 to 150 cm), and virtually all of it falls during the wet season.

Temperatures in the savanna regions are much like those in the other tropical climates: Monthly averages are above 65°F (18°C) all year. The precipitation pattern, however, results in a higher annual range of temperature than in either of the other tropical climates (10° to 20°F). During the wet season, clouds keep monthly averages in the upper 70s and lower 80s. In the dry season, monthly averages may climb into the upper 80s or low 90s.

The pattern of climate in the tropical regions is fairly predictable. The greater the degree of latitude from the equator, the longer is the dry season because with increased latitude, equatorial low pressure weakens and subtropical high pressure becomes progressively more effective. Total annual precipitation generally decreases with increased latitude, and the annual temperature range increases directly with latitude.

In the tropical regions, minor annual fluctuations in total precipitation cause little concern for local inhabitants because rainfall is, overall, more than adequate. But as the dry season increases at more interior locations, or as latitude increases, even a minor fluctuation in total precipitation may have a profound effect on the daily lives of the people in those regions. For example, the failure to receive even one-half of the average normal rainfall during the first four wet seasons in the Sahelian zone of west and central Africa in the early 1970s resulted in mass starvation of both people and animals.

The tropical climates are regions of many contrasts. Some of the world's most densely populated areas are located in tropical regions, for example, India and Indonesia. On the other hand, certain tropical climate regions rank among the least populated areas of the world, such as the Amazon basin and the Zaire basin. Although tropical regions have some of the most lush natural vegetation growth in the world, many problems confront the tropical farmer: Soils are rapidly depleted of mineral foods and organic matter; sizeable markets are often inaccessible; and mechanization, which could benefit many farmers, is much too expensive. Despite these problems, certain crops grown in tropical climates are of worldwide importance, such as rubber, coffee, cocoa, tea, sugar, and pyrethrum.

In the tropical climates, mineral resources are available in varying amounts and quality. Unfortunately, some of the largest reserves of various ores are located in highly inaccessible regions where the cost of establishing a mining operation and transporting the raw material often makes such a venture a risky one. Some mineral resources, however, are of great economic importance, particularly in some less developed countries; for example, diamonds, gold, copper, bauxite, cobalt, and tin, to name just a few.

Forests in the humid tropics contain some of the most valuable tree species in the world, such as teak, mahogany, and bamboo. Unlike the forests of the middle latitudes, where certain tree species often grow in large stands, however, tropical tree species tend to grow as isolated individuals. Consequently, to harvest these trees requires searching for them, hauling expensive equipment often considerable distances through dense vegetation growth in order to cut the tree, and then hauling the cut tree out of the forest to a milling site. Thus, despite good potential, the forest resources of the humid tropics remain largely untapped.

Raising animals for meat and milk in tropical regions has met with only limited success. Cattle and sheep are difficult to raise, except in the drier regions, due to the large number of insects that transmit a wide variety of diseases to domesticated animals. There is a very definite correlation between the presence of the tse-tse fly, for example, and the absence of cattle. Efforts are being made to promote the raising of hogs and chickens in place of cattle since they are generally more resistant to disease.

Dry Climates (B)

The dry climates, which include the *steppe climate* (BS) and the *desert climate* (BW), are characterized by their lack of precipitation. In the lower latitudes, dry climates are generally explained by the presence of subtropical high pressure; in middle and higher latitudes, dry climate regions are located in the interior of continents or on the leeward side of mountains protected from moist air masses.

Although lack of precipitation is the outstanding feature of the dry climates, annual precipitation totals by themselves do not necessarily define a dry climate—temperature must also be considered. An annual precipitation total of 30 inches (75 cm), for example, is considered a normal amount for a humid climate in the middle latitudes. But in lower latitudes where temperatures are high all year, evaporation effectively reduces the amount of usable moisture. Thus, 30 inches of rainfall in parts of the Sahelian zone of Africa might produce only a semiarid, or *steppe,* climate. Of course, there are many dry locations that receive such limited amounts of precipitation annually that even with cool temperatures, the amount is not adequate to raise food crops.

Steppe (BS) climates are located on the margins of the tropical savanna climate regions as well as along the interior margins of humid continental climates and in leeward positions in the middle latitudes (see Figure 8–2). The steppe regions receive less total precipitation than the savanna areas. Annual amounts of 10 to 20 inches (25 to 50 cm) are common. Unfortunately for farmers, annual amounts tend to vary considerably from year to year; rains can produce a bumper crop one year, and the next year a drought could completely destroy a crop.

Temperatures in the steppe regions are dependent largely on latitude and season. In subtropical steppe regions, average monthly temperatures are high year round, usually above 65°F. In the middle-latitude steppe regions, a larger annual range of temperatures is characteristic, reflecting the change in seasons. Daily maximums and minimums are similarly controlled by latitude and the seasonal variations associated with it.

The *desert climate* (BW) marks the extreme in moisture deficiency. Most of the world's largest deserts lie within the subtropical high-pressure belt within 5° or 10° of the Tropics of Cancer and Capricorn. The Sahara, Kalahari, Arabian, and Australian deserts are just a few examples (see Figure 8–2). In the middle latitudes, desert climate regions are located in the interior of continents where moist air masses do not penetrate, and along the lee side of mountains. The Gobi Desert, the Great Basin, and the Takla Makan Desert attribute their aridity to one or more of these factors.

Annual precipitation totals in desert regions are so small that any question of the effects of low or high temperatures on evaporation is merely academic. In some desert locations, years have passed without measurable amounts of rainfall; so annual averages under these conditions are meaningless. When rain does fall, it tends to be a cloudburst, often depositing several inches of rain in a few minutes. This cloudburst may be followed by no rainfall for several months or even years.

As we have seen, annual temperatures are a function of latitude. In the tropics and subtropics, temperatures can be exceptionally high all year due to the absence of water vapor and clouds. Monthly averages can easily exceed 80°F (27°C), with daily maximums reaching well over 100°F (38°C). Daily minimum temperatures can be quite cool, often dropping into the 50s. Desert climates experience the greatest daily range of temperature of any subclimate. Following daytime temperatures of more than 100°F, a night temperature of 50°F can be a chilling experience. At the middle and higher latitudes, temperatures range from the 80s or 90s in the summer months to below freezing in the winter months. Daily temperature ranges at these latitudes may be large in the summer, but winter ranges are considerably less.

Desert regions are windy places during the daytime. Great variations in the albedo of surface materials result in large temperature differences and significant local pressure gradients within short distances. Winds, as a result, tend to be strong and gusty. During evening hours, by contrast, local pressure gradients are very small. In fact, deserts are noted for their calm evenings.

Among the most intriguing phenomena in the desert is the *mirage*. What appears as water rippling on the desert surface is actually light being refracted as it passes from a layer of warm, light surface air to slightly cooler, denser air above. Local convection currents cause the mirage to appear to shimmer.

Despite the disadvantage of the lack of water, for centuries people have continued to occupy desert regions. Deserts have provided an ideal environment for many cultures and isolated groups of people who sought desert regions to isolate themselves from conflicting groups and cultures. Both the bushmen and the Mormons, for example, sought this isolation. Today, dry climates are viewed by many people as healthy places, generally free of insects and diseases. The large retirement communities in the desert southwest United States attest to this attitude.

Water is the key to future desert development. If water is available, desert soils can be made highly productive by irrigation techniques. Without a doubt, as long as water is available, people will continue to seek the desert environment for warm, dry air. The advisability of continually developing dry climatic regions, however, is certainly a question to be pondered. In a world where population continues to increase and local water shortages occur with increasing frequency, should we risk diverting water from humid regions to dryland areas? We must also consider the high energy demands that we need to be comfortable in the deserts, for example, air conditioning and refrigeration.

Humid Mesothermal Climates (C)

The climates of the lower middle latitudes are called *mesothermal,* a term that means *moderate temperatures.* There are three mesothermal climates: the Mediterranean, the humid subtropical, and the marine west coast. The mesothermal climates differ in terms of their season of heaviest precipitation and their degree of warming during the summer months.

The *Mediterranean* regions (Csa, Csb) are located between latitudes of 30° and 40° (see Figure 8–2). Mediterranean climate is controlled by subtropical high pressure and westerly winds. During the summer months, these regions are dominated by subtropical high pressure. Precipitation, as a result, is very sparse, and summer is characteristically a dry season. With the onset of winter, the subtropical high shifts equatorward, and the Mediterranean regions come under the influence of subpolar low pressure. Hence, westerly winds carry moisture onshore during this season. In North America, moisture is transported by westerly winds from the eastern Pacific, whereas it is the eastern Atlantic that provides the source of moisture for the Mediterranean regions of southern Europe. The dry summer season in the Mediterranean regions lasts from two to four months. Annual precipitation totals are not large, mostly in the 15-to-30–inch (38-to-76–cm) range. In fact, annual totals are not unlike those of many steppe climate regions. In contrast to the steppes, however, most of the precipitation falls during the cool season, and thus little precipitation is lost to evaporation. The greatest drawback to this seasonal pattern of precipitation is the lack of moisture during the peak of the growing season; so irrigation is an important part of agriculture in the Mediterranean regions of North America.

Temperatures in the Mediterranean regions reflect the lower–middle-latitude locations and the seasonal precipitation pattern of this climate. In the summer, average monthly temperatures are in the 70s and 80s, with daily maximums and minimums only slightly less extreme than in dry climatic regions. In the winter, the air is more humid and the daily range is considerably smaller. Winter monthly averages are near 40°F (4°C). The annual temperature range is 30° to 40°F. Frost is a rare occurrence except at higher elevations. Fogs, triggered by cool offshore ocean currents, are common in many areas.

The *humid subtropical* regions (Caf) lie at latitudes of 25° to 40°. Unlike the Mediterranean regions, however, which lie on the western side of a continent and experience dry conditions during the warm season, the humid subtropical regions lie on the eastern side of a continent and receive the maximum rainfall in the warm season, and thus generally have no real dry season (see Figure 8–2).

The humid subtropical climates lie in gaps between major oceanic subtropical high-pressure cells. Consequently, maritime tropical air masses penetrate these regions in both summer and winter, and they bring abundant precipitation, especially in summer. In the winter, cold polar air clashes with maritime tropical air to generate cyclonic or frontal precipitation. Average annual precipitation totals vary from region to region. The average amount is 50 inches (127 cm), but totals will vary from 30 to 100 inches (76 to 254 cm).

Temperatures vary with the season. Summer temperatures are usually high. The warmest month's average is 75° to 80°F (24° to 27°C). Combined with high humidities, high temperatures make summer hot and sultry. Even at night, temperatures remain high since the water vapor in the air retains most of the heat. The daily range, therefore, is quite small. In the winter, the coldest monthly average temperature is near 50°F (10°C). Daily maximums in winter will reach 50° to 60°F, and daily minimums are most often in the 40s. High humidities in winter often make people feel damp and chilly despite rather moderate temperatures. Frost is infrequent but can form with invasions of continental polar air masses.

The *marine west coast* regions (Cbf) are a third mesothermal climate. The marine west coast climate regions differ from the other mesothermal locations because they are located at relatively high latitudes (see Figure 8–2). Most marine west coast regions lie poleward of 40° latitude, and in western Europe they extend to 65° latitude. The moderate temperatures at these latitudes are controlled by warm offshore ocean currents. As westerly winds blow over these warm ocean currents, the winds are warmed and, in turn, transport this warmth to the adjacent coastal regions. This onshore circulation pattern is relatively constant during the year; therefore, these regions have moderate temperatures most of the year.

Monthly temperatures in the marine west coast regions average in the upper 60s during the summer and in the upper 40s in the winter. Thus, the annual range is only 20° to 30°F, which is extremely small for the higher middle latitudes. Summer daily maximums may on rare occasion reach the 90s, but more often they are in the 70s. Daily minimums in summer are usually in the 60s. In a similar fashion, daily extremes in the winter months range from 50° to 30°F. Frost is more common to this climate than to the other mesothermal regions, but it is not at all excessive considering the latitude.

Precipitation totals vary with each region, varying from 25 to 100 inches (63 to 254 cm). In North America, annual totals rank among the highest on the continent, usually exceeding 40 inches and in some places exceeding 100 inches. The precipitation is largely orographic as warm, moist air from over the ocean is forced to rise as it encounters the coastal mountains. There is a definite winter season maximum due to increased instability in the lower layers of air as they are warmed by contact with the warm ocean surface. In summer, precipitation is restricted largely to the higher elevations since ocean air masses are stable at lower layers, having been cooled by contact with the ocean surface. In western Europe, the absence of coastal mountains allows oceanic air masses to penetrate into the interior of the continent. Temperatures in western Europe have the same general range as in North America. The major difference in the two regions is that in western Europe, annual precipitation amounts are less, and at most locations, slightly more precipitation is received in the summer. Annual totals in western Europe average 30 inches (76 cm), and the precipitation is generated largely by cyclonic activity. Precipitation in both North America and western Europe is mostly in the form of rain. However, snow may fall between 10 and 15 days each winter.

Fog and clouds are very common characteristics of the marine west coast climate. Fogs are an almost daily occurrence in the coastal regions of western Europe in the winter months. Clouds cover the skies of these regions about 200 days each year. Indeed, the marine west coast regions are the cloudiest climate on earth.

Due to the moderate temperatures, the mesothermal climates are among the most popular climatic regions with people in the world. People have vacationed in the Mediterranean regions for centuries to take advantage of the clear skies and abundant sunshine. Many of the world's glamor vacation spots lie in the Mediterranean regions, such as Southern California, the French Riviera, and the Greek Islands. The Mediterranean regions are also noted for many agricultural products, such as citrus fruits, olives, grapes (wine), and vegetables.

The humid subtropical regions also benefit from moderate year-round temperatures. The lack of a dry season means that often two crops can be grown in a single year. In North America, many former cotton lands have been converted into pasture for beef cattle, while such crops as citrus fruits, tobacco, sugar cane, cotton, wheat, and barley are grown. The southern pine forests in the United States rank as the second leading lumber-producing region in the country, as well as a leader in pine-related products such as turpentine, pitch, tar, and resin. In Asia, most of the humid subtropical regions are intensively cultivated for rice. In the United States, many large industries have relocated from the heavily industrialized northern states to the southern states or California to take advantage of the cheaper energy and the ability to attract workers to these regions of desirable climate.

The marine west coast regions of North America are well suited to the lumber industry. The Douglas fir, the redwood, and other large coniferous species provide the raw material for one of the world's leading lumbering regions. Agriculture, except on a small scale, is not a major industry in this region of North America. In western Europe, on the other hand, agriculture is of prime economic importance. Root crops (such as potatoes, beets, and turnips), grains (such as winter wheat and rye), and other crops (such as deciduous fruits, berries, and grapes) dominate the western European economic landscape. The climate is also conducive to pasture and feed crops, and dairying is an important agricultural activity.

Humid Microthermal Climates (D)

The microthermal climates are located between 35° and 75° latitude in the Northern Hemisphere (see Figure 8–2). They are not a feature of the Southern Hemisphere due to the lack of large landmasses at those latitudes. The microthermal climates include the humid continental climates (Daf, Dbf) and the subarctic climates (Dcf, Ddf). The two categories of microthermal climates are differentiated by the severity of their winter season. The subarctic regions are located further poleward and more to the interior of continents than are the humid continentals, and for that reason, they experience longer, colder winters and shorter, cooler summers.

Temperatures in the *humid continental* climatic regions range from a high monthly average of 65° to 75°F, to a low monthly average of 10° to 35°F. Daily maximums in summer may exceed 100°F, and winter daily minimums may plunge to −25°F or less. The annual range of temperature reflects both the latitude of a location and the distance from coastlines. Temperature ranges increase with latitude and with increasing distance from large water bodies.

Precipitation in the humid continental regions averages 25 to 35 inches (64 to 89 cm) annually. A maximum of precipitation falls in the warm season, and no months are really dry. Most precipitation is the result of cyclonic activity. Winter precipitation is usually in the form of snow, and snow may form a permanent ground cover for a few weeks to several months. On a day-to-day basis, weather in the microthermal climates is highly variable due to the continuous conflict between polar and tropical air masses and the alternate invasion of cyclones and anticyclones as they are guided across the middle latitudes by westerly winds.

The *subarctic* regions represent the most poleward and most extreme of the microthermal climates. They are distinguished from polar climates in that the warmest average monthly temperature exceeds 50°F (10°C). In the humid continental regions, polar air masses dominate the climate in winter, and tropical air masses dominate in summer. But in the subarctic regions, polar air masses are common year round. Polar air masses are characterized by cool to cold temperatures and little precipitation.

Subarctic climates have the greatest annual range of temperature of any subclimate. The highest monthly average is generally between 50° and 60°F, and the coldest monthly average is about −60°F. Verkhoyansk, USSR, for example, has an annual temperature range of 118°F (48°C): a July average of 60°F and a January average of −58°F. Daily maximums in summer may exceed 90°F, and daily minimums in winter have reached as low as −108°F.

Cold polar air is not conducive to large amounts of precipitation. In fact, annual totals of less than 20 inches (50 cm) are very common and rival the totals in many semiarid regions. The year-round relatively low temperatures, however, reduce the loss of moisture to evaporation. Thus, subarctic regions still qualify as humid climates, and in many areas the environment does support an extensive forest cover. Precipitation is mostly cyclonic, and the majority falls as rain during the warm season. Total moisture during the winter months is small, and what does precipitate is in the form of dry snow. Snow covers the ground for as much as eight months of the year; however, total annual snowfall is usually greater in humid continental regions due to the increased moisture available in slightly warmer winter air.

The microthermal regions, especially the humid continental climates, are well suited to agriculture during the warm season. The growing season is adequate for most crops to mature, having both high temperatures and abundant precipitation during the period. The humid continental regions harbor some of the leading agricultural areas of the world. The Corn Belt of the midwestern United States, for example, is one such area. Corn, oats, soybeans, hay, winter wheat, and barley are all grown successfully. In the United States a large portion of the corn crop supports an extensive meat-raising industry of beef cattle, hogs, and sheep. In the more poleward sections of the humid continental regions, where growing seasons are slightly shorter and closer to larger urban centers, dairying is common. Root crops such as sugar beets and potatoes replace grains as the primary crops in the cooler sections of these climatic regions.

In the subarctic regions, growing seasons are generally too short for most grains to mature. Also, potential frosts in summer restrict the growing of many vegetables. Only certain root crops and frost-resistant vegetables are grown with any appreciable success. The large stretches of coniferous forest that cover much of these regions supply pulp and paper industries. Trappers, hunters, fishermen, and miners have had some success in subarctic regions.

Polar Climates (E)

The polar climates include those regions where the warmest average monthly temperature does not exceed 50°F (10°C). The tundra climate (ET) and the icecap climate (EF) comprise the polar climates. Latitude is the key to these regions (see Figure 8–2). Except for some coastal locations, most polar climates are poleward of 75° latitude. Usually only two to four months have average temperatures above freezing, and killing frost is likely at any time.

In the tundra climatic regions, average summer temperatures range from 35° to 50°F. In the winter, average temperatures vary by region. In Siberia, for example, January temperatures average near −35°F, while in North America, north central Canada averages −25°F, and Labrador, −10°F. Daily maximums in summer can reach into the 80s, with daily minimums generally in the 40s. Maximum daily temperatures in winter are usually only slightly above 0°F, and daily minimums can drop to −80°F.

Precipitation in the tundra regions rarely exceeds 12 inches (30 cm) annually, even in coastal locations. Polar high pressure and cold air limits the formation of precipitation. Precipitation is at a maximum in summer due to increased cyclonic activity. Normally, precipitation in summer is in the form of rain, but it can snow even in mid-summer. The extremely cold air in winter does not yield much snow. What snow does accumulate, however, may not melt for up to 10 months. In fact, the snow is so dry and in such small amounts, that as much as 80% of tundra regions have exposed bare ground.

While tundra regions occupy mostly coastal locations, *icecap* climate regions are located more toward the interior of continents and at slightly higher latitudes (see Figure 8–2). Icecap regions are the most extreme, continuously cold climates on earth, for they are too far poleward to be invaded by warm air masses. No average monthly temperature is above freezing. Even in summer, temperatures are cold since the sun never rises very high in the sky and nearly 80% of the sun's energy is reflected by a permanent snow and ice surface. Average summer temperatures in Greenland are near +12°F, while in Antarctica the temperature is closer to −10°F, despite the fact the sun shines almost 24 hours a day. In winter, monthly averages in Greenland are about −50° to −60°F, and in Antarctica they are closer to −60° to −80°F. Daily summer maximums may reach 50°F in Greenland and 40°F in Antarctica, but any daily maximum above 32°F is unusual. Daily minimums even in summer are usually below 0°F. A record daily minimum of −127°F was recorded in Antarctica in the winter of 1958.

With low year-round temperatures, it is understandable that precipitation amounts are also low. Most stations record less than 3 to 4 inches (7 to 10 cm) annually. Precipitation is in the form of occasional dry snow, even in summer, and some accumulation is derived from frost deposits. Snow cover is permanent all year.

Human use of the polar regions is restricted to specialized economic activities such as fishing and hunting. People must adapt all phases of their lives to the harsh conditions of these climates. Clothing, shelter, transportation, and food all require specialized adaptations in order for people to live comfortably in this environment. The rigors of the polar climates and the expense of adapting to these environmental conditions explain the absence of large populations in these regions.

Highland Climates (H)

The highland climates (H) include a wide range of climatic conditions, each of which is basically a modification of a climate found at the same approximate latitude but at a lower elevation. The degree of modification in climates due to altitude depends on the amount of elevation change and the degree of exposure to the sun. In effect, there is no such thing as a typical highland climate.

Below 4000 feet (13,200 m) the changes in climate are not very noticeable, but above 6000 feet (19,800 m) the changes become more distinct. Highlands generally receive more insolation than do lower elevations because the atmosphere is less dense at higher altitudes. Since the air is less dense, radiation escapes rapidly from highlands at night. The net result is a large daily range of temperature.

Temperatures in highland regions decrease with increased altitude at the rate of about 3.5°F/1000 feet (6°C/1 km). In tropical latitudes, the lower temperatures of highlands provide refuge from the hot summer for people who inhabit lower elevations. Each zone of elevation in the tropical highland is suited to particular agricultural activities (see Figure 8–3). In the middle and higher latitudes, highland regions do not hold the same attraction as they do in tropical areas because temperatures are generally not very high, even in summer at lower elevations. Therefore, human use of highlands at middle and higher latitudes is limited.

Seasonal variations in temperature in highland regions parallel the seasonal variations at similar latitudes and lower elevations. Of greater variation than temperature between highlands and lowlands is the change in precipitation amount with elevation. In tropical highlands precipitation increases to

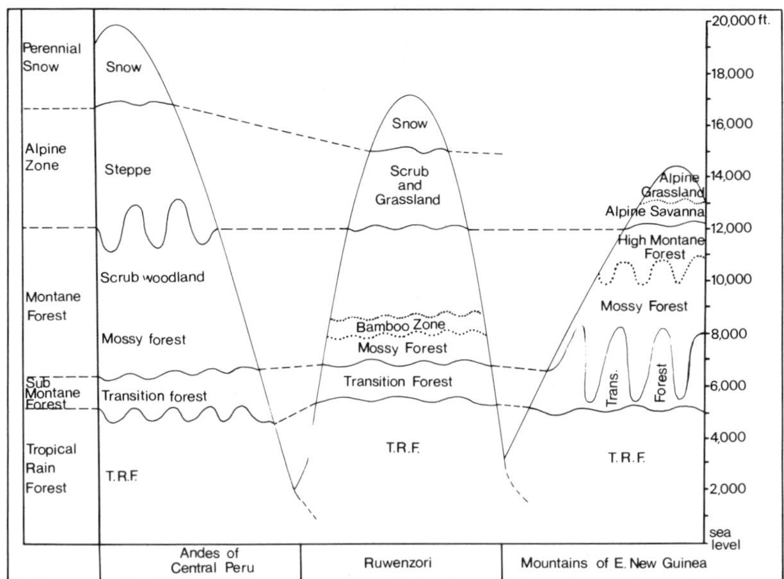

Figure 8–3

about 5000 feet (16,500 m), above which elevation precipitation amounts decrease. In the middle and higher latitudes, however, precipitation tends to increase steadily to the top of the tallest mountains. Precipitation in highlands can be of vital importance in those regions where the adjacent lowlands receive only marginal precipitation. Rain and melting snow in the highlands can supply valuable water for human consumption and irrigation at lower elevations. Some highlands receive enough precipitation to supply water to feed large rivers and support dense areas of population. Both the Colorado River and the Nile River are fed by mountain streams, and both supply water to large numbers of people in semiarid and arid lowland regions.

Additional Material

Figure 8–4 gives the climograph for each of the 13 world climatic regions.

Singapore   Af
Calicut   Am
Acapulco   Aw
Monterrey   BSh
Cairo   BWh
Sacramento   Csa
°F
P (in.)
J F M A M J J A S O N D
80
60
40
20
0
-20
-40
-60
30
25
20
15
10
5

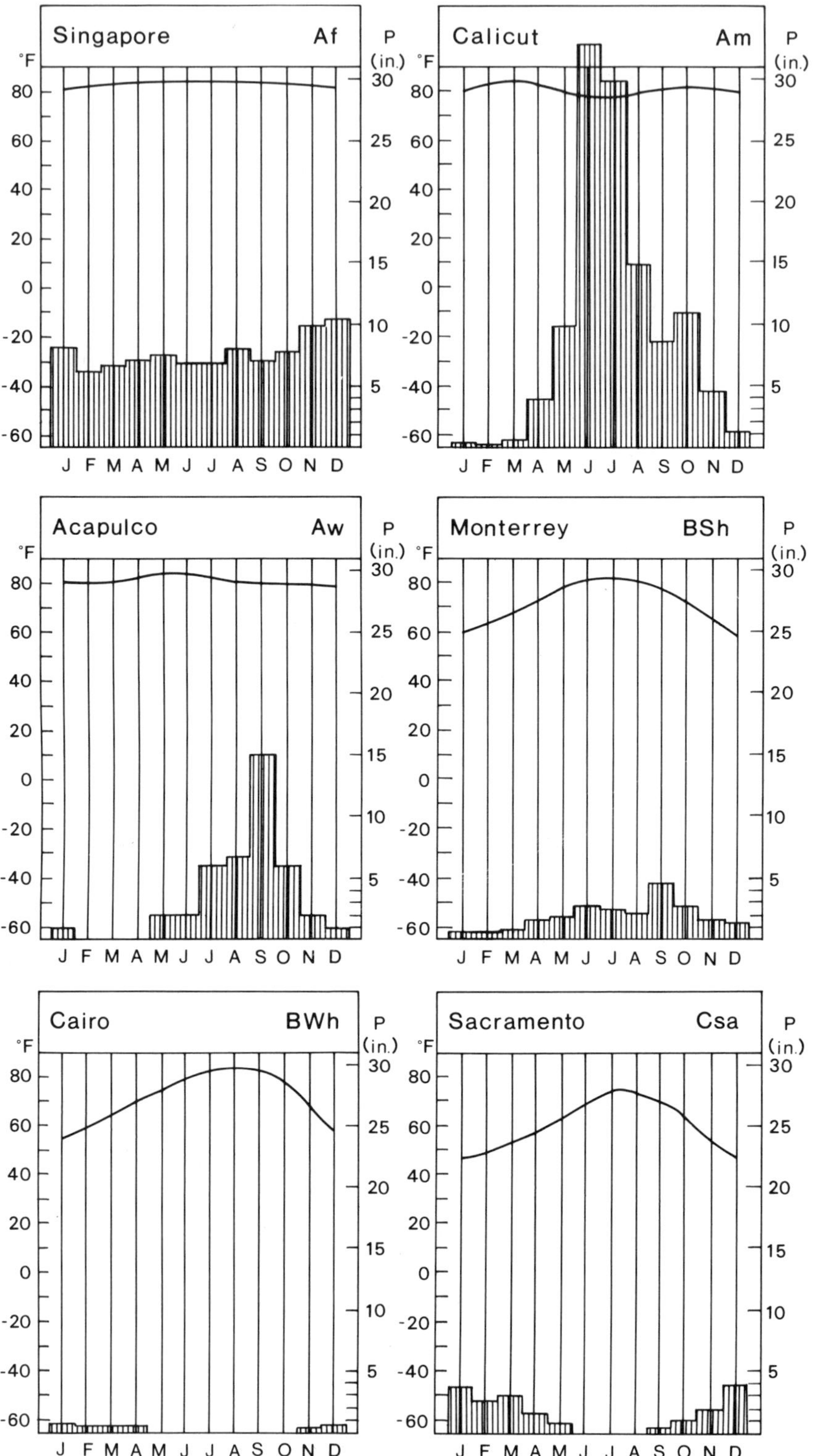

Figure 8–4

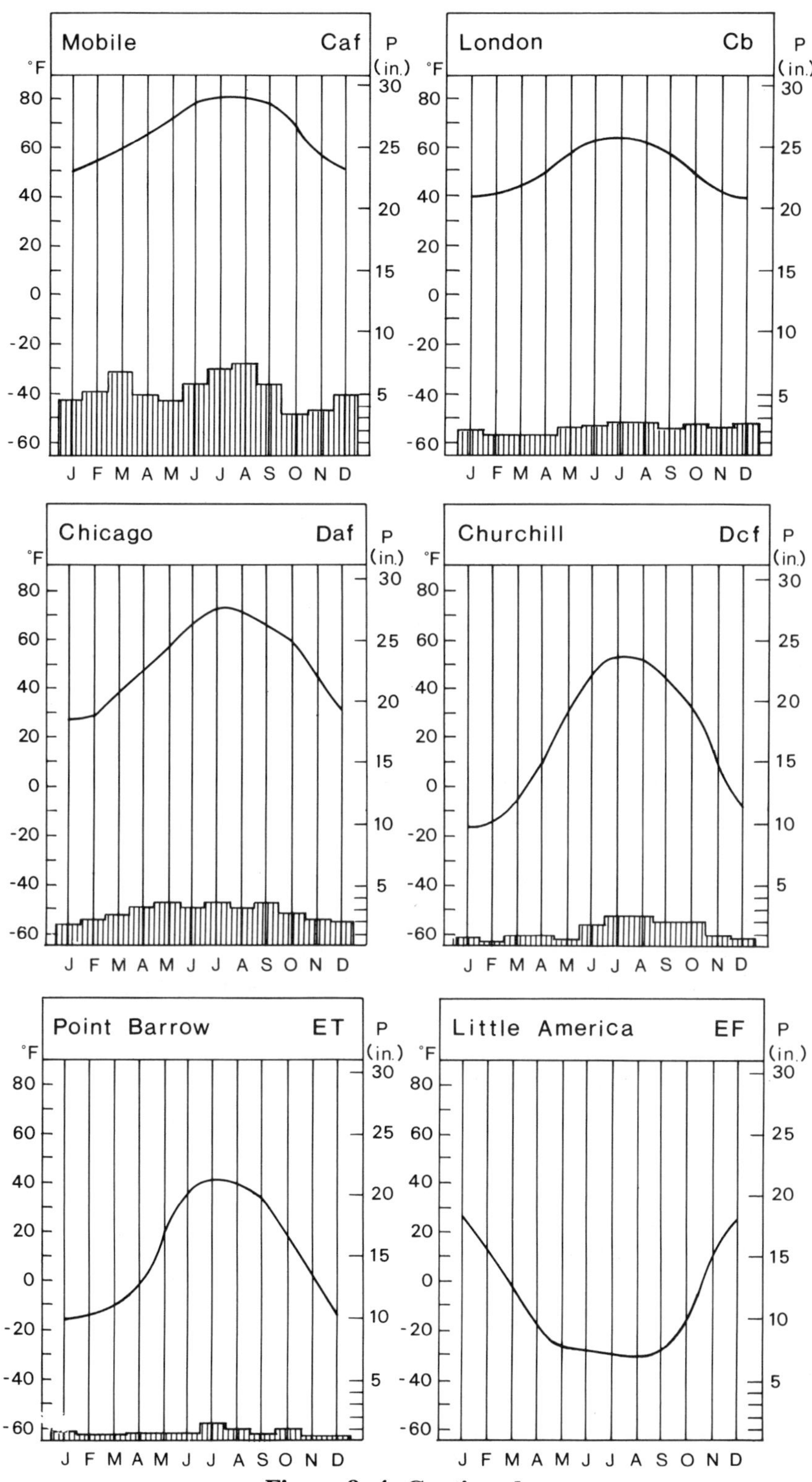

Figure 8–4, Continued

141

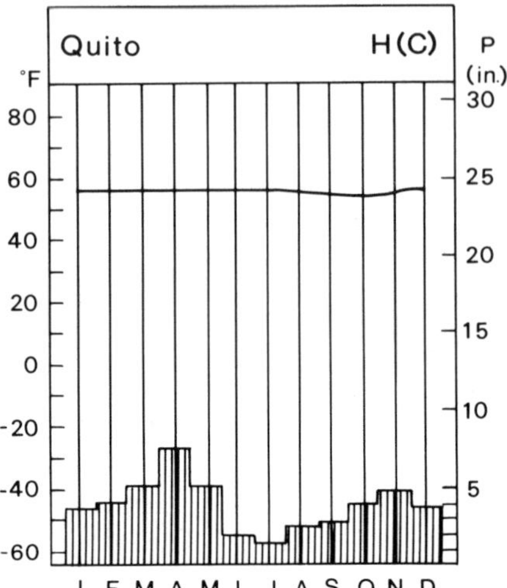

Figure 8–4, Continued

Questions

1. What are the advantages and disadvantages of a classification system?

2. If you examine the various economic maps in an atlas, you will note that the use of the land in areas of the same climatic classification may vary from one region to the next. What factors explain these variations?

3. What accounts for the higher average yearly temperatures at many places near 30° latitude than places within 5° of the equator?

4. How does the ITCZ shift during the year? Which climatic regions does it partly influence?

5. How do you explain the fact that the daily range of temperature exceeds the annual range in tropical rainforest regions?

6. How do you explain the change in the seasonal distribution of precipitation in the tropical rainforest, tropical monsoon, and tropical savanna regions?

7. How do you explain the heavy precipitation along east coasts at latitudes of 10° to 25°? What climatic region is there? What climatic region do you find in the interior of land areas at the same latitudes?

8. What factors are responsible for causing desert conditions?

9. What is responsible for strong winds during the day but calm nights in desert regions?

10. What explains the great daily range in temperature in desert regions?

11. How do you explain the existence of humid subtropical climatic regions at latitudes that are generally dominated by high pressure?

12. What explains the great variation in area covered by marine west coast climate in North America as compared to western Europe?

13. Mediterranean climates have annual precipitation totals similar to those of tropical steppe regions. What seasonal factor qualifies the Mediterranean as a humid climate?

14. What air masses affect the humid continental climate regions? How do these air masses vary with season?

15. What explains the great annual temperature range in subarctic climatic regions?

16. How do the annual precipitation totals in the icecap regions compare to the annual totals in tropical deserts? How does evaporation compare in these two areas?

17. How do temperature and precipitation totals vary with increasing elevation in tropical latitudes? In middle latitudes?

Suggested Readings

ANTHES, R. et al. 1975. *The atmosphere*. Columbus, OH: Charles E. Merrill. Chapters 10 and 11.

BATTAN, L. J. 1974. *Weather*. Englewood Cliffs, NJ: Prentice-Hall. Chapter 7.

BOUCHER, L. 1975. *Global climate*. New York: John Wiley & Sons.

GATES, D. M. 1972. *Man and his environment: climate*. New York: Harper & Row.

LOWRY, W. P. 1967. The climate of cities. *Scientific American,* August 1967. Reprint 1215. San Francisco: W. H. Freeman.

MATHER, J. R. 1974. *Climatology*. New York: McGraw-Hill.

McBOYLE, G. 1973. *Climate in review*. Boston: Houghton-Mifflin.

POWERS, E., and WITT, J. 1972. *Traveling weatherwise in the USA*. New York: Dodd, Mead & Co.

TREWARTHA, G. T. 1966. *The earth's problem climates*. Madison: University of Wisconsin Press.

————. 1968. *Introduction to climate*. 4th ed. New York: McGraw-Hill. Chapters 7–12.

Module
9

Soil Properties
and Processes

INTRODUCTION

The role of soils as part of the natural environment is a variable one. For example, both the construction engineer and the farmer view soils differently. The engineer is concerned with the thickness and support qualities of the soil compared to those of solid rock for building purposes; the farmer wishes to know the ability of the soil to grow crops. Both of these people would likely have a different definition for soil. For our purposes we will define *soil* in a traditional fashion: *Soil is a natural complex of organic and mineral matter suited for vegetation growth.* Of most importance in the definition are the words *vegetation growth*. Individual plants such as mosses and lichens may be

145

able to grow on hard rock surfaces, but to support a nearly continuous cover of larger plants requires soil.

Most people would agree that soils are anything but dynamic, but in effect, soils are extremely dynamic. That is, they are constantly undergoing changes, and although these changes are rarely detected by the human eye, the changes that do occur are of great importance to plants and therefore people.

Soil is one of the most basic resources on earth next to air and water. It is an essential link in the human food supply. The increased demand for food by ever-expanding population has brought a greater awareness of the importance of soil conservation. The soil resource must be maintained both in quality and in quantity if we hope to maintain a constant supply of food, which even today is substandard when measured in terms of the average number of calories per person per day. It has been estimated that just to stay even with current levels of consumption, food production will have to double to meet the expected population increase by the year 2000.

Soils are an integral part of the physical environment. They are closely associated with climate and vegetation variation, with the configuration of the earth's landforms, with inorganic materials, and with the biosphere, which includes humans and all lower forms of living organisms. Even a slight change related to one or more of these associated factors can be enough to greatly alter the character of a soil. Because of the relationship of soil to the other facets of the physical environment, the basic properties and processes of soil are introduced in a physical geography course.

OBJECTIVES

By the end of this module, you should be able to do the following:

1. Explain the physical characteristics of soil.
2. Explain the chemical characteristics of soil.
3. Describe the basic characteristics of each horizon of a soil profile.
4. Explain the relationship of soil to the soil-forming factors.
5. Explain the basic properties from which soil profiles develop.

KEY TERMS

texture	eluviation
loam	illuviation
structure	soil profile

consistency   soil horizon

porosity   solum

permeability   podzolization

humus   calcification

parent material   laterization

leaching   capillary action

Now you are ready to begin the audiovisual portion of this module. Select the MEDIAPAK 9 component(s) and proceed. Following is a topical outline of the audiovisual sequence. You will find this outline helpful for reference and review. After completing MEDIAPAK 9, return to this book to perform the exercises.

OUTLINE

Soil properties
> *Physical properties*
>> Texture
>> Structure
>> Consistency
>> Porosity and permeability
>> Color
>> Water
>
> *Chemical properties*
>> Nutrients
>> Acidity and alkalinity
>> Leaching
>
> *Soil profile*
>> Horizons

Soil-forming factors

Soil processes
> *Laterization*
> *Podzolization*
> *Calcification*

Exercises

1. Which of the following are terms that describe soil structure?
   A. Sand, silt, clay
   B. Loose, hard, plastic, sticky, friable
   C. Acidic, alkaline, neutral
   D. Platy, prismlike, blocklike, spheroidal

2. On the average, a soil is composed of four basic constituents in the following proportions:
   A. Water 25%, air 25%, mineral 45%, organic 5%.
   B. Water 30%, air 5%, mineral 45%, organic 20%.
   C. Water 25%, air 25%, mineral 25%, organic 25%.
   D. Water 25%, air 35%, mineral 25%, organic 15%.

3. Which soil texture particle is most likely the smallest?
   A. Silt
   B. Loam
   C. Clay
   D. Sand

4. Which form of soil water is most useful for plants?
   A. Capillary water
   B. Hygroscopic water
   C. Gravitational water

5. Which of the following is *not* a macronutrient in soil?
   A. Calcium
   B. Zinc
   C. Potassium
   D. Nitrogen
   E. Phosphorus

6. The soils of humid climate regions tend to be
   A. acidic.
   B. alkaline.

7. By leaching activity, soil materials are often removed from the A_2 horizon. This zone in the soil is termed the
   A. capillary zone.
   B. zone of illuviation.
   C. zone of calcification.
   D. zone of acidity.
   E. zone of eluviation.

8. Which of the following soil-forming factors is most capable of causing a rapid change in soils?
   A. Climate
   B. Parent material
   C. Time

      D.   Topography
      E.   Humans

 9. The laterization process is most active in
      A.   warm, humid climates.
      B.   dry climates.
      C.   temperate, humid climates.
      D.   cold, dry climates.
      E.   cold, humid climates.

10. Capillary action is a characteristic of
      A.   podzolization.
      B.   laterization.
      C.   calcification.

Summary

Soils are a dynamic part of the natural environment because they continually adjust to changes in the environment through alterations of one or more physical and/or chemical characteristics. Soil is composed of air, water, minerals, and organic matter. On the average, 45% of the soil is mineral, 25% air, 25% water, and 5% organic matter. However, few, if any, soils are "average." In fact, it is the variation in soil composition and the relationship between its composition and agricultural potential that make the science of soils so fascinating.

Physical Characteristics

To the casual observer, the physical differences in soils are most apparent. Variations of soil texture, structure, consistency, porosity, permeability, water, color, and profile are all characteristics that can generally be seen with the human eye.

Soil **texture** is the relative proportion of the various soil particles, which are, in order of largest to smallest, sand, silt, and clay. Most soils are a mixture of all three particles, in which case the texture is described as a **loam.** Should a loam be composed of mostly one texture size, such as sand, the soil texture is termed a *sandy loam*. Texture is important because it determines the amount of pore space available to hold air and water. Too sandy a soil forms too many large pore spaces, with the result that water passes through the soil too rapidly to be utilized by plant roots. A soil with an overabundance of clay, on the other hand, presents the opposite extreme, that is, not enough pore space for air or water. Generally, therefore, a soil texture that is neither too sandy nor too clayey is better suited for most plant types.

In most soils the individual soil particles combine into larger clumps, called *peds,* which give soils a particular structure. Soil **structure** is the way in which the soil particles combine together to form secondary particles (peds), which assume a distinctive pattern. Soil structure is really a measure of how well a soil can be worked by agricultural machinery. It has a direct effect on soil porosity and permeability (defined below), which in turn affect the movement of nutrients to plant roots. Soil structures are classified into four basic forms: platy, prismlike, blocklike, and spheroidal. Blocklike and prismlike soil structures have moderate rates of water infiltration, and platy and spheroidal structures generally have slow infiltration rates. Some soils lack any structure at all and are called *single-grain* soils. Single-grain soils have rapid infiltration rates. Soil structures can be changed by adding fertilizers to the soil, and by plowing, cultivating, and irrigating.

Closely associated with soil texture and structure is soil **consistency,** the resistance of soil to breaking apart. Descriptions of consistency are usually related to the moisture content of the soil. *Loose, hard, plastic, sticky,* and *friable* are just a few terms defining consistency. For most agricultural activities, a friable consistency is perferred because the soil has a nearly optimum ratio of texture particles and a workable structure.

Soil **porosity** is the proportion of air space in the soil. It provides an indication of the total amount of water that can be retained in a soil. The rate at which water, gas, or plant roots penetrate the soil is a measure of soil **permeability.** Permeability and porosity are closely related. The total amount of water that infiltrates the soil and the rate at which it does so vary both with the amount of available pore space (texture) and with the arrangement of the pore spaces in the soil (structure). Soils that are neither too sandy nor too clayey are best suited for agriculture because pore spaces are of proper amount and size to permit water to permeate the soil at a satisfactory rate for most plants. On the other hand, sandy soils are highly desirable for septic tank drainage fields because of their high rates of permeability.

One of the most obvious characteristics of soil is color. Soil color varies with moisture content, amount of organic matter, mineral content, and depth below the surface. Unfortunately, soil colors can lead to false interpretations of soil content because similar colors may be the result of widely dissimilar factors. Nevertheless, at the basic level, some generalizations about soil colors can be listed: A black or dark brown soil is often the result of abundant organic mater; a reddish or yellowish soil usually possesses a significant iron and/or aluminum content; silica produces a grayish color, and soluble salts a whitish color; a bluish soil indicates a lack of oxygen; and a mottled or spotted soil is the result of changes in environment on a seasonal basis. Most people assume that a dark soil is a fertile soil, and it often is. But a dark soil may also be infertile due to a large concentration of alkalies.

Soil water is a complex chemical solution that consists of many soluble chemical elements. It is essential to plant life because it contains the chemical elements, or nutrients, for plant growth. The availability of soil water to plant roots—the key to plant growth—is based on the following varying factors: the amount of soil water available, the manner in which it is held in the soil, and the length of time it is available. There are three forms of soil water: hygroscopic, gravitational, and capillary. Capillary water is of most value to plants because it is available in sufficient supply for an adequate length of time, and it is held in the soil in such a way that it can easily be absorbed by plant roots. Hygroscopic water, by contrast, is bonded too tightly to soil particles to be absorbed by plants, and gravitational water passes through the soil too rapidly for the majority of it to be utilized.

Chemical Characteristics

The **parent material** from which a soil is derived is unconsolidated and more or less *chemically weathered* mineral and organic matter. Parent materials for soils include mineral and rock fragments, residual material from decaying plant life, and material that has been transported and deposited by water, ice, wind, or gravity. Nearly all natural elements compose parent materials or are part of the atmosphere. Natural elements are, therefore, also a part of soil.

The chemical features of a soil are more important for plant growth than are the physical features. By chemical reaction in the soil, nutrients, or food, are made available to plant roots. The chemical elements needed by plants fall into two categories: macronutrients and micronutrients. Macronutrients are those chemical elements that are required by plants in large amounts; these include calcium, potassium, nitrogen, and phosphorus. Micronutrients are also necessary for plant growth but in lesser amounts; for example, boron, copper, iron, zinc, and manganese are just a few.

The interaction of chemical processes in soil is complex. The result of the processes is measurable in terms of the departure of the soil from chemical neutrality toward either acidity or alkalinity. The degree of acidity or alkalinity in a soil helps to determine the availability of nutrients.

---

pH

pH is a measure of the hydrogen ion activity in a soil. The more active the hydrogen ions, the more acid is the soil. A scale of acidity and alkalinity, called the *pH scale,* indicates the degree to which a soil departs from neutrality. The scale varies in value from 1 to 14, with a value of 7 an indication of a neutral soil. Soils with values less than 7 are progressively more acid with each lower number. Alkalinity increases with each successive number above 7.

Plants absorb chemical elements in solution. When the soil is too alkaline, soil water has little ability to dissolve minerals and release available nutrients. Thus, plants will not receive nutrients even though they may be present in the soil. Most soils, however, are slightly acidic since alkalies are not particularly abundant in most soils and since decomposing organic matter usually releases acid. Too much soil acidity is not good for plants, however, because certain elements may be too rapidly dissolved by an acidic soil-water solution, and other elements that do not easily dissolve may become too abundant and therefore may be toxic to plants. Most agricultural crops grow best in soils that are mildly acidic. As a generalization, dryland areas have alkaline soils, and humid areas acid soil. This is often the case because alkaline materials in dry regions are not readily dissolved by water, and the predominant form of decaying vegetation is grass, which is less acid than the decaying trees that are more common to humid regions. Acidic soils can be neutralized by applications of lime. Alkaline soils are neutralized by heavy flushing with irrigation water followed by applications of sulfur.

Leaching is a process whereby nutrients are dissolved by water percolating through the soil. The leached nutrients are transported from the upper layers of the soil to lower layers. In humid regions, leaching can be a major problem as surface layers in the soil are leached free of nutrients and thus rendered infertile. The leaching process is partially offset by **capillary action** which draws moisture and dissolved nutrients up in the soil by evaporation at the surface. Distinct layers of the soil are particularly affected by leaching. One zone in the soil from which soil materials are removed by leaching is termed the **zone of eluviation.** Soil materials removed from the zone of eluviation are deposited at a lower layer in the soil called the **zone of illuviation.** Generally speaking, west of the 100° west meridian in the United States, capillary action exceeds leaching activity. Therefore, soils in the western half of the country are relatively fertile but lack adequate moisture. In the eastern half of the United States, on the other hand, leaching is very common, and soils must be replenished by chemical fertilizers or must be allowed to remain fallow for a period of time in order to allow the supply of nutrients to replenish naturally.

Variations in soils from place to place are found not only in the surface layer but also in layers below the surface. Most soils can be subdivided with depth into distinguishable layers called **horizons,** although often only the skilled soil scientist can detect any changes from one horizon to another.

Many soil science investigations involve comparisons of **soil profiles,** cross sections of soils showing the horizons. Soil scientists label each horizon in the soil profile with the letters A, B, and C, starting from the surface and moving down. Minor variations within a given horizon are designated *subhorizons,* and numbers are attached to the horizon letter; A_2, for example, is a subhorizon of the A horizon (see Figure 9–1). The A and B horizons in the

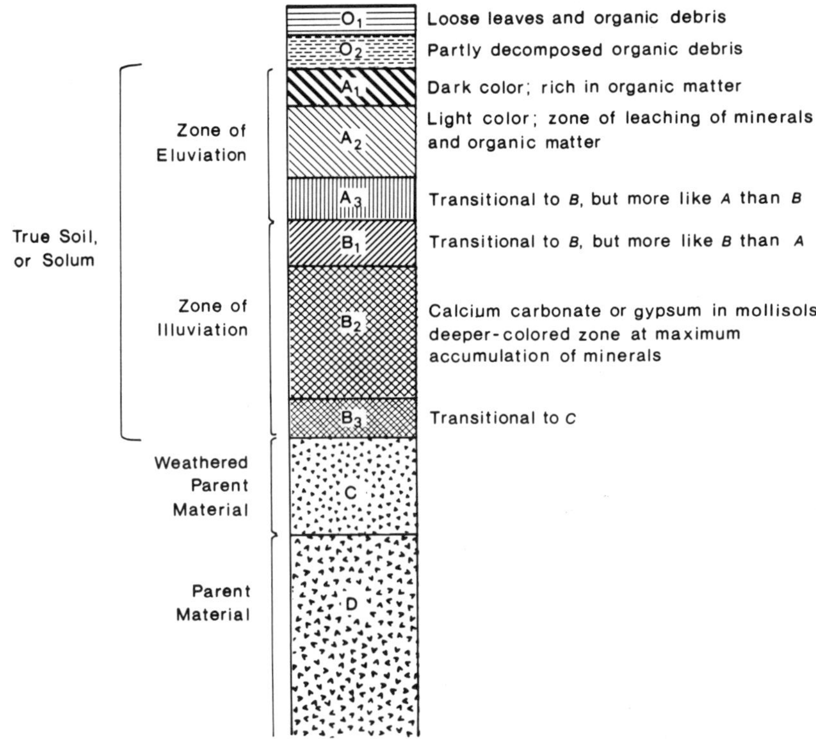

Figure 9–1

soil are the most important in terms of plant growth because they contain the bulk of the nutrients for plants. These two horizons together comprise the **solum,** or growth zone. The C horizon contains weathered parent material but no organic matter. Below the C horizon is the D horizon, which is the parent material from which the surface layers developed.

Because of the great variety of conditions under which soils develop, no two soils have exactly the same profile. Thus, by analyzing the differences in profiles, scientists can identify soil types. In one soil, a given horizon or subhorizon may be completely absent from a profile, while in a different environment, that horizon or subhorizon may be well developed.

Soil-Forming Factors

The characteristics of a soil and the rate at which those characteristics develop depend on the soil-forming factors. Soil-forming factors include parent material, climate, topography, time, the biosphere, and humans.

The *parent material* provides the raw material for a soil; it is acted upon by the other soil-forming factors. The influence of *climate* on soils is a function primarily of temperature, precipitation, and evaporation, each of which directly affects the rate of chemical reactions, the rate at which parent material breaks down, and the nature of the soil-water solution. Climate indirectly affects soils by its influence on vegetation and microorganisms. The correlation between climate and soil types is so close that the world pattern of soils and climates is very similar (see Figures 9–2A and 9–2B).

Topography refers to the physical features of a region, particularly the degree of slope and the amount of relief (difference in elevation between the tops of hills and bottom of valleys). The slope of the land and the direction in which the slopes face affect soil development. Steep slopes are generally better drained than more gentle slopes. As a result, less water penetrates soils on steeper slopes and thus soil development is inhibited. The soil on slopes tends to be thin and poorly developed, and if the slopes are extremely steep, a mature soil with distinct horizons may not form. Erosion is more active on steeper slopes, while on more gentle slopes, the rate of erosion is generally matched by the rate of soil formation. On the other hand, valley bottoms and flat lands often have very deep soils, but many are poorly drained, thus preventing adequate aeration for plant growth.

The direction in which slopes face, or *slope orientation,* is particularly important to soils in the middle and higher latitudes. In the Northern Hemisphere, north-facing slopes have colder temperatures and are generally wetter than slopes with a southerly exposure. If south-facing slopes are not too steep, they tend to be better suited to agriculture than north-facing slopes at these latitudes. The influence of slope orientation on soils also is indirectly affected by the degree of vegetation development on different slopes.

The amount of *time* necessary for a soil to develop from parent material and reach maturity varies considerably due to the variable rates of weathering and the varied effects of climate, topography, and organic matter on soil formation. A soil is mature if it has well-developed soil horizons and is in balance with its environment. In other words, a soil is mature if it has generally maintained its present profile for a considerable length of time and is not undergoing major adjustments within horizons. Time as a soil-forming factor is difficult to measure. The longer a soil has to develop, the less is the influence of its parent material and the greater are the influences of climate and vegetation. Generally, it requires hundreds or thousands of years for a soil to mature. On the average, it requires 1000 years to form 1 inch of topsoil, although many topsoils can form in less time.

The *biosphere* includes both plant and animal life, both of which affect soil formation. Vegetation acts to hold soils in place by reducing erosion. It also influences the soil water budget by its effect on infiltration and transporation. When vegetation dies, it provides organic nutrients to the soil. Such

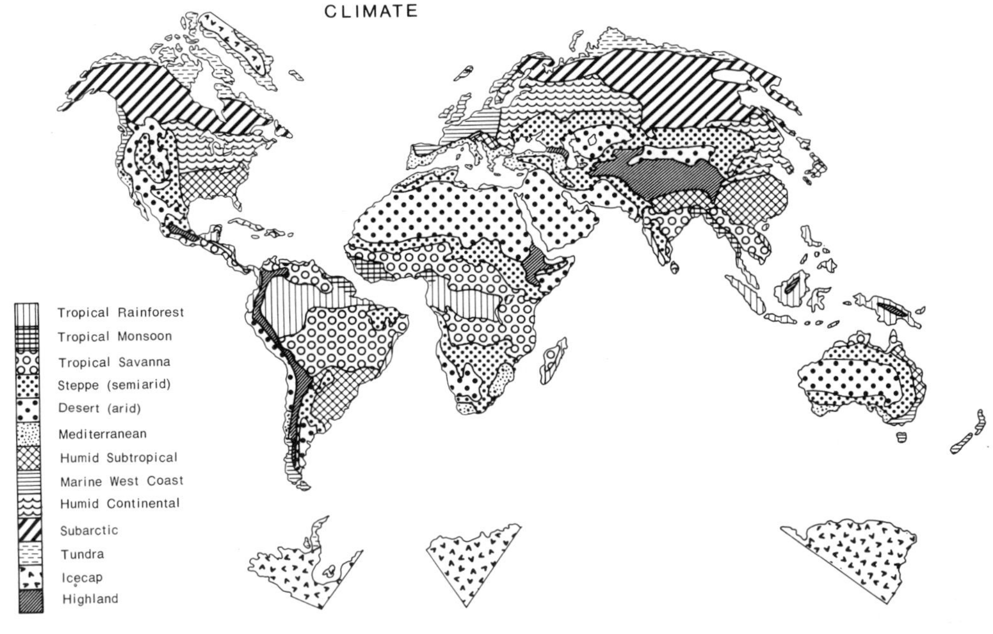

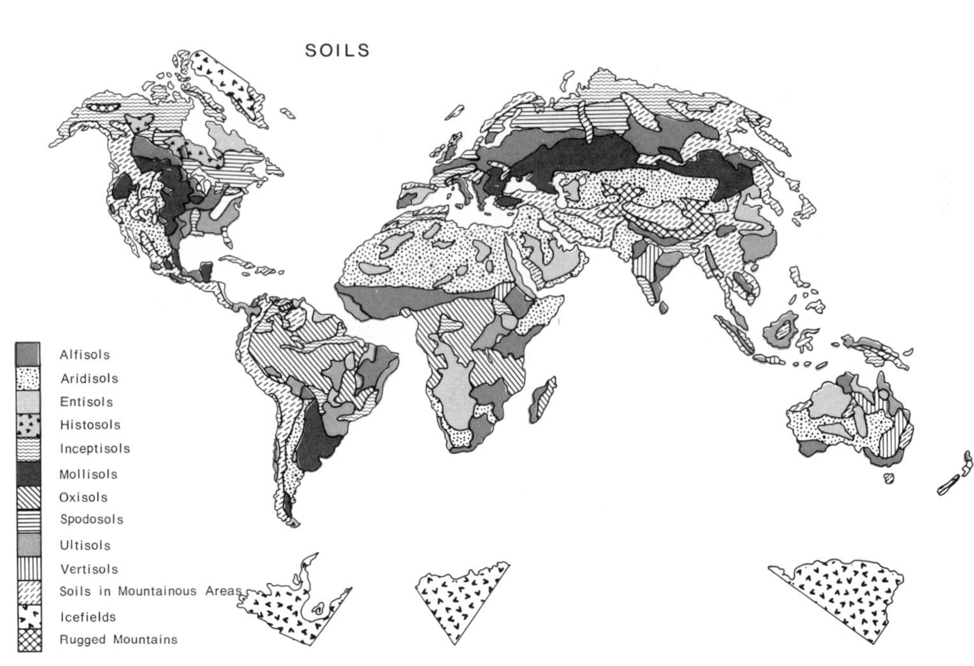

Figure 9–2

155

decayed organic matter in soil, called **humus,** supplies the food for microorganisms that in turn assist in preparing nutrients for plants. Larger animal life, such as earthworms, ants, slugs, and termites, assist in mixing the mineral and organic compounds in the soil. This mixing action by animal life can also result in changes of texture, structure, and chemical quality in the soil.

Humans are part of the biosphere, but they are listed separately as a soil-forming factor because of the profound changes that they can make in soils in just a very short period of time. While it may take thousands of years to form a topsoil, humans can completely destroy it in minutes with a bulldozer. People affect soils by changing the land. For example, plowing, raising crops, fertilizing, and irrigating are just a few activities by which we affect soils. We are capable of maintaining and even improving soils with proper conservation, but too often we have been guilty of destroying the soil by depleting it of its fertility and accelerating erosion.

Soil Processes

Soils develop under three general processes that are functions primarily of climate: laterization, podzolization, and calcification.

Laterization is a process common to soils in humid tropical and subtropical climates. Laterized soils develop under conditions of high temperatures and abundant moisture. Under these conditions, rocks, minerals, and humus are broken down rapidly. Leaching is intense, and therefore laterized soils are characterized by generally low fertility with heavy concentrations of iron and aluminum in the upper soil horizons because these elements are the most difficult to dissolve. The iron and aluminum provide the characteristic red and yellow color of these soils. Often the leaching process is so intense and operates at such depth that, in effect, only the A horizon is found within 20 to 30 feet of the surface. The laterization process continues year round in tropical climate regions due to the absence of cool seasons.

Podzolization is a soil-forming process that is common in humid climates with seasonal temperatures. The process is best developed in the higher middle latitudes where the summers are short and cool and the winters are long and severe. As in the humid tropical regions, leaching is also active in the humid sections of the middle latitudes. Unlike the laterized soils, however, podzolized soils are less intensively leached because there are cooler temperatures and less moisture and also because the process does not work at all during the winter months when the soil is frozen. Podzolized soils tend to possess distinct A and B horizons and several distinct subhorizons. Most podzolized soils have definite zones of eluviation and illuviation. Soil color varies with subhorizons; but the upper A horizon tends to be dark brown, the eluviated lower A horizon grayish, and the illuviated B horizon dark brown.

Both podzolized and laterized soils require applications of fertilizers if they are to be continually utilized for agriculture.

Calcification differs from the other soil processes because it develops soils in dry climate regions. Leaching and eluviation are relatively unimportant in calcified soils. Instead, the general motion of soil water is toward the surface due to capillary action. The result of capillary action is often a concentration of alkaline minerals in the upper layers of the soil. The concentration of alkaline minerals, such as calcium carbonate, in the upper layers of the soil is not a problem for agriculture unless the concentration is very near the surface where it can be overabundant, and hence toxic to plants. In regions where calcification is the primary soil process, grass vegetation often provides large quantities of organic matter for humus. Consequently, some of the calcified soils that develop under grass vegetation rank among the most fertile soils in the world. The prairie soils of Illinois, Iowa, Kansas, and Nebraska are representative examples.

Questions

1. What disadvantages do soils that are too sandy or too clayey have?

2. How do porosity and permeability vary with soil texture?

3. What factors are generally responsible for red, yellow, black, and bluish colors in a soil?

4. In what forms is water held in the soil? How do the forms differ in terms of plant utilization?

5. What are some parent materials for soil?

6. What is the difference between a macronutrient and a micronutrient? What are the main macronutrients?

7. What is important about the degree of acidity or alkalinity in a soil? What problems occur if soil is too acidic or too alkaline? How can extremely acidic or alkaline soils be neutralized?

8. What is the effect of leaching on soils? What is the relationship of leaching to eluviation and illuviation zones in the soil?

9. What is the difference between leaching and capillary action? In what general climatic regions is each most active?

10. How does climate affect soil formation?

11. What factors of slope orientation can affect soil development?

12. What is the importance of humus and microorganisms in the soil?

13. How do humans affect soil development?

14. How do laterization and podzolization differ?

15. How does calcification differ from the processes of humid regions?

Suggested Readings

BUNTING, B. T. 1965. *The geography of soil.* London: Hutchinson & Co.

DONAHUE, R.; SHICKLUNA, J.; and ROBERTSON, L. 1971. *Soils.* 3d ed. Englewood Cliffs, NJ: Prentice-Hall.

GREENLAND, D., and DE BLIJ, H. J. 1977. *The earth in profile: a physical geography.* San Francisco: Canfield Press. Chapters 10 and 12.

KELLOGG, C. E. 1950. Soil. *Scientific American,* July 1950. Reprint 821. San Francisco: W. H. Freeman.

McNEIL, M. 1964. Lateritic soils. *Scientific American,* November 1964. Reprint 870. San Francisco: W. H. Freeman.

STEILA, D. 1976. *The geography of soils.* Englewood Cliffs, NJ: Prentice-Hall.

VAN RIPER, J. E. 1971. *Man's physical world.* 2d ed. New York: McGraw-Hill. Chapters 15 and 16.

Module 10

World Soil Patterns

INTRODUCTION

As we saw in the last module, soils vary in composition from place to place; in fact, no two soils are exactly alike. Many soils, however, develop in the same type of environment and have roughly the same basic characteristics. Like climatic regions, therefore, soils are classified into separate groups, with all members of a given group possessing the same general characteristics. By classifying soils into a group, soil scientists can make comparisons between and unravel some of the complexities in soils around the world without being overwhelmed by the infinite number of minor soil variations.

Soil classification dates back as far as 400 years ago when the Chinese identified different soils. More recently, in the late nineteenth century, a classification was developed by a Russian scientist, V. Dokuchaiev. The classification of Dokuchaiev was based on the knowledge that soils develop and change due to factors other than simple alterations of parent material. That is, soils are considered as part of an entire system that includes climate, the biosphere, vegetation, landform, and time. The Russian classification became the model for a later classification developed in 1914 by C. F. Marbut, chief of the U.S. Soil Survey. The Marbut classification was revised several times and reached its final form in 1936. That classification grouped soils basically by their genetic characteristics. The Marbut system of so-called great soil groups is still presented in many introductory physical geography and soil textbooks. After World War II, however, it became apparent to the professional soil scientist that the Marbut classification was not satisfactory for classifying all soils, particularly those which had been greatly altered by erosion or made from recently deposited materials. During the 1950s, the U.S. Department of Agriculture began to develop a new classification that would be responsive to all soil types.

The U.S. Department of Agriculture uses the U.S. Comprehensive Soil Classification System to classify soils. This system classifies soils based on surface horizon or subhorizon properties that are easily identified and can be quantified or measured. The Comprehensive Classification consists of 10 major soil orders with 10,000 separate soils for the United States alone. This system is now widely accepted in the United States, but it still has not been endorsed by other countries.

OBJECTIVES

By the end of this module, you should be able to do the following:

1. Describe the primary characteristics of each of the 10 orders of the U.S. Comprehensive Soil Classification System.
2. Describe the general world locations of each of the 10 orders of the Comprehensive Classification.
3. Describe the general agricultural value of each of the 10 orders of the Comprehensive Classification.

KEY TERMS

The basic terminology for soils is given in Module 9.

Now you are ready to begin the audiovisual portion of this module. Select the MEDIAPAK 10 *component(s) and proceed. Following is a topical outline of the audiovisual sequence. You will find this outline helpful for reference and review. After completing* MEDIAPAK 10, *return to this book to perform the exercises.*

OUTLINE

U.S. Comprehensive Soil Classification
Diagnostic characteristics
Ten soil orders
Entisol
Vertisol
Inceptisol
Aridisol
Mollisol
Spodosol
Alfisol
Ultisol
Oxisol
Histosol

Exercises

1. Which of the following soil orders can be found in almost every climatic region,
    and includes thick sand deposits and alluvial deposits?
    A. Ultisol
    B. Vertisol
    C. Aridisol
    D. Entisol
    E. Histosol

2. Which soil order occupies the most total land area?
    A. Oxisol
    B. Aridisol
    C. Inceptisol
    D. Spodosol
    E. Vertisol

3. The _______________________ soil order forms under podzolization, sup-
    ports forest regions, and has a coarse-textured, acidic upper A horizon.
    A. vertisol
    B. histosol
    C. alfisol
    D. oxisol
    E. spodosol

4. Which soil order is located both in the middle latitudes and in the subtropical
    regions, and has a distinct clay horizon in the B horizon?
    A. Alfisol
    B. Spodosol
    C. Inceptisol
    D. Oxisol
    E. Histosol

5. Ultisols develop under the _______________________ process.
    A. laterization
    B. podzolization
    C. calcification

6. Which soil order is characterized by a subsurface horizon that is about 1 foot
    thick and that consists of iron and aluminum oxides?
    A. Vertisol
    B. Histosol
    C. Oxisol
    D. Mollisol
    E. Alfisol

7. Which of the following soil orders includes soils that are more than 50% organic
    matter?

    A. Vertisol
    B. Inceptisol
    C. Mollisol
    D. Spodosol
    E. Histosol

8. Which soil order covers an extensive portion of the Great Plains?
    A. Oxisol
    B. Spodosol
    C. Mollisol
    D. Aridisol
    E. Alfisol

9. Which soil order is utilized primarily for its bunch grass vegetation which supports grazing activities?
    A. Spodosol
    B. Inceptisol
    C. Mollisol
    D. Aridisol
    E. Oxisol

10. Which soil order includes tropical soils that are able to absorb and retain fertilizers, but become extremely heavy and sticky when wet?
    A. Oxisol
    B. Vertisol
    C. Histosol
    D. Millisol
    E. Spodosol

Summary

The U.S. Department of Agriculture uses the U.S. Comprehensive Soil Classification to classify soils. This system classifies soils based on surface horizon or subhorizon properties that are easily identified and can be quantified or measured. The Comprehensive Soil Classification identifies 10 major soil orders into which all soils can be classified. Each of the soil orders includes soils with similar properties. Because of their similarities, all soils within a given order are assumed to have developed in the same manner. All the order names end with the suffix *sol,* which means soil. The prefix for each order is either a descriptive term or a Latin derivative. The 10 orders and the derivation of their prefixes are listed in Table 10–1.

Table 10–1

Soil Order	Root	Root Meaning
Entisol	*e*, *n*, and *t* from word *recent*	
Vertisol	*Verto*	To turn
Inceptisol	*Inceptum*	Inception, or beginning
Aridisol	*Aridus*	Arid
Mollisol	*Mollis*	Soft
Spodosol	*Spodos*	Wood ash
Alfisol	*Al* and *Fe*	Aluminum and iron
Ultisol	*Ultimos*	Ultimate
Oxisol	*Oxide*	Iron and aluminum oxides
Histosol	*Histos*	Tissue

Entisols

Entisol soils lack any well-developed profile either because there was insufficient time for a profile to mature or because the unique character of the physical environment under which they formed prevented such a development. Included among the entisols are soils that are newly exposed by erosion on steep slopes; alluvial, or stream-deposited, sediments; fresh glacial deposits; some marsh soils that are especially waterlogged; and thick sand deposits, as in dune regions. Entisols comprise 12.5% of the world's soils, and they can be found in almost every climatic region on the earth (see Figure 10–1).

Entisols vary in the quality of their agricultural potential. Some entisols are inadequately drained; others are too dry or too stony; and still others, such as alluvial soils, rank among the most fertile in the world.

Vertisols

Vertisols also lack well-developed profiles. These soils have large concentrations of expanding clay. When the clay is moist, it expands and the soil heaves upward slightly. When the clay dries out, it contracts, and the soil may settle at an angle. Constant heaving and contraction can actually cause the soil to turn over. At the surface, cracks may appear in excess of 1 inch in width and extend to a depth of 20 inches (50 cm) or more.

Vertisols are best developed in those climatic regions that have a distinct wet and dry season, for example, in savanna and Mediterranean regions (see

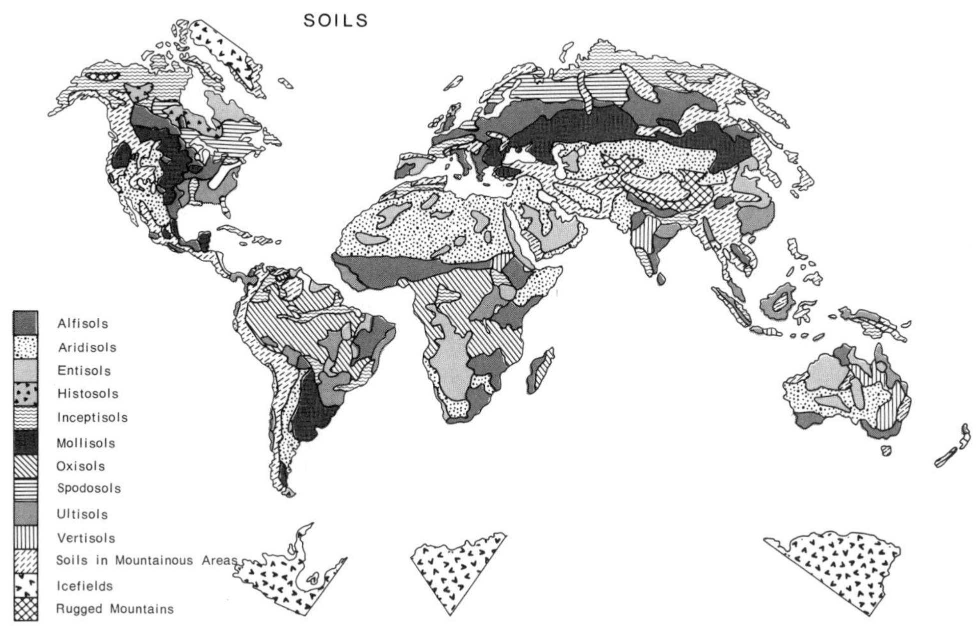

Figure 10–1

Figure 10–1). Vertisols include 2.1% of the world soils. They are among the better soils for agriculture in tropical regions because they absorb and retain fertilizers very well. On the other hand, they can be very difficult to plow since the clay can be very heavy and sticky when wet. Also, when they dry out and crack, they are highly susceptible to wind erosion.

Inceptisols

Inceptisols are a third order of soils that lack good profile development. These soils form from a wide variety of parent materials and under widely different environments. Generally, however, all inceptisols are subject to the removal of some elements by leaching. Inceptisols are located in tropical and subtropical climates and in tundra regions where the subsoil is permanently frozen or is supersaturated with water (see Figure 10–1).

Entisols, vertisols, and inceptisols all rank as *youthful* soils. Each order lacks profile development, and in each order, water control is a key factor for agriculture. Flooding and poor drainage are often a cause for crop failure on these soils, although some success has been achieved growing cotton, soybeans, rice, sugar cane, and vegetables. The overall lack of a stable natural vegetation cover as well as a poorly developed soil structure makes each of these soil orders susceptible to rapid erosion and too low in humus content.

Still, some of these soils are very fertile, such as the blacklands of Texas and Alabama and the alluvial soils of China and Egypt.

The remaining soil orders include soils that have developed a mature profile under a relatively stable vegetation and climate environment.

Aridisols

Aridisols develop where moisture is deficient. Thus, they usually have shallow horizons and a relatively high concentration of soluble minerals and coarse surface textures. These characteristics are caused by the lack of chemical weathering and leaching that results from the lack of water. The lack of water restricts the amount of vegetation cover, which explains the lack of humus in these soils.

Aridisols are usually characterized by a distinct layer of calcium carbonate in a subhorizon of the A or B horizon, or they have a concentration of mineral salts in these horizons. Aridisols can be highly productive soils provided that adequate irrigation water is available and humus is added. Agriculture that depends on irrigation can intensify salt and mineral concentrations in the soil's upper horizons, and it is very expensive. Therefore, aridisol soils are often used to support grassland to be used for grazing animals rather than tilled for planting crops. Aridisols occupy more total area than any other soil order, or 19.2%. The location of aridisols is closely correlated with the regions of desert and drier steppe climates around the world (see Figure 10–1).

Mollisols

On the wet margins of the steppe climatic regions are the mollisol soils. Mollisol soils are deeply weathered, and they have thick horizons and dark coloration due to larger amounts of humus and moisture. They are closely associated with grass vegetation. Indeed, the most striking feature of these soils is their very dark brown or black coloration in the upper horizons.

Dark coloration in mollisols is a product of both the combined growth and decay cycles of grass and the local moisture distribution. Grasses have fibrous root systems, and thus when they decay, large amounts of humus are supplied to the soil along with abundant nutrients. Mollisols vary in fertility based on the degree of grass cover and the amount of moisture available. As the vegetation cover increases along with increased moisture, the soils become progressively darker and more fertile. Where tall prairie grass is the vegetation cover, some of the most productive soils for agriculture can be found.

Mollisols are generally excellent soils for large-scale commercial grain farming and livestock grazing. Wheat, rye, barley, corn, and sorghum are all grown on mollisols, with climate determining the importance of each crop. Wheat, for example, requires a cool, moist period during its early growth stage, and a hot, dry harvest period. Corn, on the other hand, requires considerably more moisture than does wheat. Consequently, corn is grown only along the wettest margins of steppe climatic regions. Rye and barley are grown further poleward since they can tolerate lower temperatures. Sorghum is a very drought-resistant crop and hence is grown with wheat along the dry margins of steppe climates. The most frequent problem facing farmers who till mollisol soils is the threat of drought.

Mollisols occupy 9% of the world soils and are concentrated in the Great Plains of North America, the Pampas of Argentina, and the steppe climatic regions of central Eurasia (see Figure 10–1).

Spodosols

Spodosols are associated with the forested sections of the subarctic climatic regions. As shown on Figure 10–1, they include the northeastern portions of North America and the northwestern portions of Europe. Spodosols comprise 5.4% of the world soils.

Spodosols develop in large part under needleleaf evergreen trees. These trees prevent most sunlight from reaching the soil surface. As a result, soil temperatures are low, and the rate of microbacterial action is retarded. Thus, organic matter is decomposed very slowly, and soil water is generally highly acidic due to the acids released from the decomposing pine needles. Consequently, leaching of one or more of the upper subhorizons is often intense and forms a distinct zone of eluviation, usually in the A horizon. The upper A horizon is usually coarse-textured and acidic; the B horizon becomes packed with more finely textured particles derived from the A horizon, and it has relatively abundant amounts of iron and aluminum (zone of illuviation).

Leaching of the upper subhorizons, plus cool temperatures, restricts agriculture on spodosols. Most often the soils are used to support the growth of needleleaf evergreen trees used for lumber and pulp and paper. Some crops are grown on spodosols, such as oats, wheat, hay, and corn for silage. For spodosols to be productive, however, large quantities of lime and fertilizers are required. Root crops, such as potatoes and sugar beets, are grown successfully in some regions, but they too require particular kinds of fertilizers.

Alfisols

In many areas, spodosols are closely associated with alfisols. Spodosols develop from parent materials abundant in sand, while alfisols are clay-dominated soils. Alfisols are characterized by generally light-colored surface horizons and a distinct clay layer, usually in the B horizon. The clay layer is formed as fine clay particles are leached out of the A horizon and are deposited in the B horizon.

Alfisols are located in both the middle latitudes and subtropical regions (see Figure 10–1). They are the second largest group of soils, comprising 14.7% of the earth's soils. No other soil order exists in as many diverse climatic and vegetation environments as do the alfisols. They range from moist to seasonally dry climates, and from under broad-leaved, deciduous trees to thornbush and savanna grass.

In the middle latitudes, alfisols develop under broad-leaved, deciduous trees. As a result of both a slightly more moderate climate and the lack of acidic pine needles, alfisols are generally more fertile than spodosols. The slightly warmer climates allow organic matter to decompose, maintaining a supply of humus in the soil and giving the upper horizons a dark coloration. Although clay layers do form in the middle latitudes, frost action during the winter helps break up the layer somewhat so that water can percolate through it. In subtropical areas, the clay layers tend to be very thick and often impervious to water. Such impervious clay layers are called *clay hard pans,* and during wet seasons, subtropical alfisol soils are often poorly drained.

The great variety of climatic conditions under which alfisols develop leads to a multitude of agricultural uses for these soils. In the middle latitudes, and particularly the United States, alfisols support some of the most intensive forms of agriculture. Corn, oats, soybeans, and alfalfa all thrive on alfisol soils. Cash grain farming, dairying, and livestock raising are all common activities. A large portion of the Corn Belt is on alfisols. To maintain fertility in the soils, fertilizers, particularly nitrogen, phosphorus, and potassium, must be supplied at regular intervals. In subtropical locations, uses of alfisols vary from subsistence agriculture to livestock grazing.

Ultisols

Ultisols are similar to alfisols in that they possess a clay layer in the B horizon. Unlike alfisols, however, ultisols have relatively infertile A horizons, and the entire profile is developed to a greater depth. In many instances, the close similarity of ultisols to alfisols has some soil scientists convinced that only time differentiates these two soil orders. In North America, for example, the alfisols located in the southern Great Lakes states were covered by glacial deposits during the Pleistocene age, while ultisols were not. Consequently, ultisols have been weathered for a considerably longer period of time than have the alfisols. Given sufficient time, the alfisols may evolve into ultisols in these regions.

Ultisols are located in humid subtropical and humid tropical climate regions (see Figure 10–1). With warm temperatures and abundant moisture, leaching and eluviation of soluble minerals are intense in the A horizon. Thus, the A horizon has a yellow brown color, and the illuviated B horizon develops an argillic (clay) layer highly concentrated with iron oxides, thus giving the B horizon a reddish yellow color. Often the B horizon contains an impermeable layer composed of not only clay but also other texture particles; this layer is called a *fragipan*. Ultisols comprise 8.5% of the world soils.

Ultisols have a wide range of uses. Some crops that do not grow well farther poleward, such as cotton and peanuts, do very well in these climatic regions where growing seasons are long. If fertilizers are applied, such crops as corn, oats, and tobacco grow well. In the southeastern United States, needleleaf evergreens are raised on ultisol soils to provide the raw materials for extensive pulp, plywood, and pine-products industries.

Oxisols

Oxisols are located in regions of humid tropical climate (see Figure 10–1). They comprise 9.2% of the world soils and are easy to classify because they are the only soils that possess an oxic horizon. Oxic horizons are subsurface horizons at least 12 inches (30 cm) thick that consist of a mixture of iron and aluminum oxides. The iron and aluminum oxides are abundant because the more soluble minerals are removed by intense leaching. As a result, the soil subhorizons have a distinct red and/or yellow color. Oxic horizons may also contain concentrated accumulations of clay and quartz sand that may harden into a claypan.

Tropical broad-leaved evergreen trees provide the vegetation cover for oxisol soils. Despite abundant organic matter on the forest floor, little humus enters the soil because abundant microorganisms ingest the organic material before much can enter the soil system. Low organic matter and intensively leached upper horizons, therefore, make most oxisols infertile. Consequently, intensive fertilization is required to make these soils more than of subsistence quality for agriculture. Certain plantation crops, such as rubber and cocoa, do well on these soils because they have relatively low nutrient demands.

Histosols

Histosols are the least extensive of the soil orders. They make up less than 1% of the world soils. Unlike the other soil orders, which are primarily mineral soils, histosols are mostly organic. That is, more than one-half their volume is composed of organic matter. Histosols are usually saturated or nearly saturated with water year round. These soils are often referred to as *bog, moor, peat,* or *muck* soils. They can form in virtually any climatic region, even in arid regions where drainage is impeded.

The deposition of organic matter into a water-saturated environment rapidly leads to the reduction of oxygen in the wet subsurface by oxygen-demanding microorganisms. The reduction of oxygen retards the rate of decomposition of the organic matter. The results are a steady accumulation of organic matter and a thickening of the organic layer of the soil.

Histosols can be used for intensive forms of crop production provided that the amount of water is closely managed. When partly drained, histosols can be highly suitable for various fruit and vegetable crops such as cabbage, carrots, celery, cranberries, beets, and potatoes. To insure maximum yields, farmers must supply phosphorus and potassium and prevent the soil from drying out. Should the soil dry out, it would become highly susceptible to fire.

Questions

1. Why do entisols lack a well-developed profile? What is meant by a *youthful* soil? Which of the 10 orders would you classify as youthful?

2. What accounts for the great extent of aridisols around the world?

3. What are the problems of agriculture on aridisols?

4. Under what type of vegetation do mollisols develop? What is significant about the relationship between mollisols and their vegetation?

5. How do needleleaf evergreen trees contribute to the character of the soil profiles of spodosols?

6. How would you compare the spodosol soils with the middle-latitude alfisols in terms of their physical characteristics and agricultural potential?

7. Why do oxic horizons develop in oxisol soils?

8. What are the problems of agriculture on oxisol soils?

9. What is the agricultural potential of histosol soils?

Suggested Readings

Refer to the Suggested Readings section of Module 9 for additional soil references.

EYRE, S. R. 1963. *Vegetation and soils.* Chicago: Aldine Press.

Module
11

Vegetation Controls

INTRODUCTION

There are a great variety of vegetation forms around the world; these greatly diverse forms reflect the climatic and soil characteristics of a given region. For example, the extremely tall trees and abundant variety of plant species common to the tropical rainforest regions are not found in the middle or higher latitudes. At the same time, species native to the high-latitude tundra regions differ completely from species native to tropical deserts. Because vegetation is a function of climate and environment, geographers, like botanists, are interested in plant species, but they want to know why a plant species grows at a particular location. Instead of analyzing the internal makeup of a plant, as botanists do, geographers analyze the environment in which the plant grows—the climate, the soils, and the topography, and also the effect of us humans on the total environment.

The kinds of vegetation that grow in a particular region can provide evidence to scientists not only about the present environment but also about the past. The growth pattern of tree rings, for example, can tell dendrologists (tree scientists) what the climate was like as many as several hundred years ago. In addition, the layering of pollen types found in corings of mud taken from the bottom of lakes can indicate to scientists what types of plants grew in a given area and when they grew there. As a result, deductions can be made about the past climates of that area. Thus, vegetation reveals the nature of past and present environments and is, for that reason, an essential part of the study of physical geography.

OBJECTIVES

By the end of this module, you should be able to do the following:

1. Explain the relationship between plant succession and climax vegetation.
2. Explain why plant life-forms usually grow at distinct layers.
3. Explain the importance of moisture for plant growth.
4. List the categories of plants based on their water requirements.
5. Explain the importance of sunlight to plants.
6. Explain the importance of soils to plants.
7. Describe the effects of slope on plant life-forms.
8. Explain vegetation change with altitude.

KEY TERMS

plant community	xerophytes
plant succession	hydrophytes
climax vegetation	mesophytes
plant life-forms	photosynthesis
deciduous	vertical zonation
coniferous	timber line
evergreen	

Now you are ready to begin the audiovisual portion of this module. Select the MEDIAPAK 11 *component(s) and proceed. Following is a topical outline of the audiovisual sequence. You will find this outline helpful for reference and review. After completing* MEDIAPAK 11, *return to this book to perform the exercises.*

OUTLINE

Plant communities
Development and change

Plant life-forms

Factors of vegetation variations
Climate
Water
Temperature
Sunlight
Soils
Topography
Elevation
Humans

Exercises

1. The distribution of plant communities is related most directly to
    A. topography.
    B. soil.
    C. climate.
    D. elevation.
    E. water.

2. The vertical layers of plant life-forms are the result of differing degrees of need
    from the environment.
    A. True
    B. False

3. Which of the following is a characteristic of a needleleaf tree?
    A. Small surface area of needles reduces transpiration.
    B. Large number of needles for photosynthesis.
    C. Deep root system for drawing moisture from sandy soils.
    D. A and B above.
    E. All of the above.

4. Forest species in humid continental regions are generally classified as
    A. mesophytes.
    B. hydrophytes.
    C. xerophytes.
    D. parasites.

5. The formation of vertical layers of vegetation due to competition for light is
    termed
    A. plant succession.
    B. plant life-forms.
    C. plant stature.
    D. vertical zonation.
    E. germination.

6. Soils may trigger a change in vegetation, and vegetation may trigger a change in
    soils.
    A. True
    B. False

7. Which of the following are factors of vegetation growth based on topography?
    A. Excessive drainage
    B. Saturated soils
    C. Thin and stony soils
    D. More or less sunlight
    E. All of the above

8. Tundra can be a climax vegetation.
    A. True
    B. False

9. The final result of plant succession is a
    A. plant life-form.
    B. vertical zonation.
    C. climax vegetation.
    D. stable climate.
    E. tropical rainforest.

10. Which factor is most likely to disrupt vegetation over a short period of time?
    A. Climate
    B. Topography
    C. Soils
    D. Elevation
    E. Humans

Summary

Plant Communities and Life-Forms

Plants live in distinct groupings, called **plant communities,** each of which contains members that have about the same ecological requirements. There are four major types of plant communities: forests, grasslands, desert shrubs, and tundra. Each plant community represents a very broad classification of plants, and climate plays the key role in determining the location of each community.

The climate of a region may be altered with time, and in response, the plant communities that exist in the region may change. Some plants will flourish under the new environment, while others will die off. The development of a new plant community in response to an alteration in environment is called **plant succession.** Should the environment stabilize for an extended period, plant succession will evolve to the ultimate form—the **climax vegetation.** When a climax vegetation cover has been produced, the environment contains the greatest combination of plant life that it is capable of supporting. A climax vegetation can be in the form of any of the major plant communities (see Figure 11–1).

Climax vegetation, sometimes called the *natural vegetation,* is rare because of continuous interruptions to the environment by human activities. Thus, most plant communities are in continuous change. Nevertheless, analyzing natural vegetation communities improves our knowledge of the interrelated nature of the physical environment within a region.

Within a plant community, there is a definite hierarchy of **plant life-forms,** which are categories of plants based on the form into which the plants

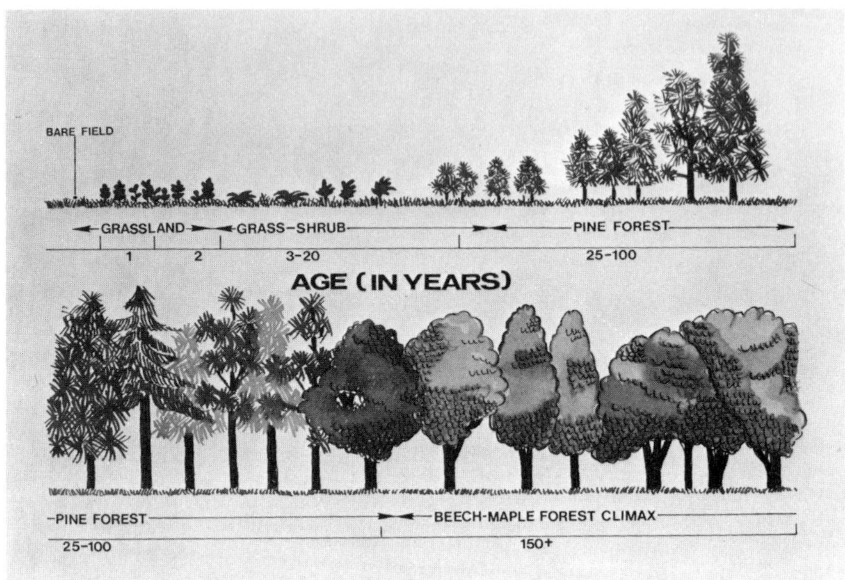

Figure 11–1

grow. The categories are based on such factors as the size and stratification of the plants, the degree to which they cover the ground, and their leaf forms. The four basic categories are trees, shrubs, herbs (including grasses), and mosses. The largest life-forms are the dominant members of a community in terms of the demand on the environment, and they form the uppermost level of vegetation above the ground. The smaller life-forms grow closer to the ground and are of smaller stature. Consequently, plant communities are organized into layers or tiers, with each layer composed of similar life-forms.

Factors of Vegetation Variations

Natural vegetation varies from place to place due to differences in *climate, soil,* and *landform.* Let's discuss *climate* first. Water is essential to vegetation because it is involved in all plant processes, including germination, growth, and reproduction. The size of a plant is often a determinant of the amount of water that it consumes and transpires. For example, large trees require large amounts of water, whereas mosses have very low moisture requirements. It is not by accident that some of the largest trees in the world are found in the tropical rainforest or in the Pacific Northwest of the United States, where annual moisture totals are high. Leaf dropping is also a response to changing moisture requirements by plants, as is the nature of the root system a plant sends out. Plants can be classified by their water requirements: (1) **xerophytes**

A.

B.

C.

Figure 11–2

require very little moisture (see Figure 11–2A); (2) **hydrophytes** require continuous moisture (see Figure 11–2B); (3) and **mesophytes** require moisture during specific stages of their life cycle (see Figure 11–2C).

All plants have an optimum temperature for each life process. Generally, the greater the seasonal variations of moisture and temperature, the greater must be the adaptability of the plants. Closely related to temperature requirements are light requirements. Plants require light to carry on **photosynthesis,** the food-making process in green plants. Light regulates the rate of photosynthesis. Plants may suffer from either insufficient light or excess light. Many plants adjust to seasonal or even daily variations in light. In middle and higher latitudes, south-facing slopes receive much more light than do north-facing slopes. Consequently, vegetation often grows in greater profusion and to a greater height on south-facing slopes. On the other hand, certain tree species grow better on the cooler, moister north-facing slopes farther poleward.

Soils have a variable impact on vegetation. At times, in fact, it is difficult to determine whether the soil characteristics result from or determine the type of vegetation that can grow there. In some cases, the type of soil may be absolutely essential to the vegetation, while in other places it is mainly the climate or the landform that is important, not the soil. Thus, the relationship between soils and vegetation must be assessed separately in each region before a final judgment can be made.

Topography, or the shape of the land surface, may influence vegetation growth by the degree and the orientation of the slope. Slopes that are excessively steep do not provide good footing for large plant life-forms, and they are usually lacking in soil moisture due to their thinness and rapid drainage. On the other hand, flatlands in humid regions often experience just the opposite problem; that is, too much moisture which creates a constantly saturated soil. (Slope orientation was discussed above.)

Elevation of the land can also affect vegetation communities; particularly in tropical highlands. As temperature delines with increased altitude, the progressively cooler environment favors certain types of plants over other types. As a result, a definite pattern is established, with certain types of vegetation growing within certain elevation limits. In addition, by influencing sunlight, moisture, winds, and soils, highland regions also affect vegetation.

Questions

1. If you were seeking some natural vegetation around the world, where would you go to look for it?

2. What factors might cause a significant change in a region's vegetation cover?

3. What are the four basic plant life-forms?

4. What is the importance of available moisture to plant growth?

5. What is the relationship between moisture and leaf loss?

6. What characteristics of needleleaf trees make them more tolerant than deciduous trees to seasonal moisture changes?

7. Into what classes can vegetation be listed based on its moisture requirements? Give an example for each class.

8. Why are there so many species of plants in the tropical rainforest but progessively fewer species as latitude increases?

9. What are the advantages and disadvantages of slope orientation toward the sun?

10. What changes in soils might initiate a change in the vegetation on those soils?

11. What is the effect of increased elevation on vegetation? Where might increased elevation be an advantage for vegetation growth? A disadvantage?

Suggested Readings

BAKER, H. G. 1970. *Plants and civilization.* 2d ed. Belmont, CA: Wadsworth.

EYRE, S. R. 1963. *Vegetation and soils.* 2d ed. Chicago: Aldine Press.

KREBS, C. J. 1972. *Ecology: the experimental analysis of distribution and abundance.* New York: Harper & Row.

LAUBENFELS, D. J. 1970. *A geography of plants and animals.* Dubuque, IA: William C. Brown.

MONEY, D. C. 1965. *Climate, soils, and vegetation.* London: University Tutorial Press.

POLUNIN, N. V. 1960. *Introduction to plant geography and some related sciences.* New York: McGraw-Hill.

WATTS, D. 1971. *Principles of biogeography.* New York: McGraw-Hill.

Module
12

World
Vegetation
Patterns

INTRODUCTION

Like the world distributions of climate and soils, the world distribution of
vegetation would be impossible to describe without basic categories based on
large-scale variations. Since changes in vegetation are largely functions of
climatic change, climatic characteristics, particularly moisture and temperature, are the most logical factors to delimit the boundaries for vegetation
classes. Indeed, if you compare the world vegetation map (Figure 12–1) with

181

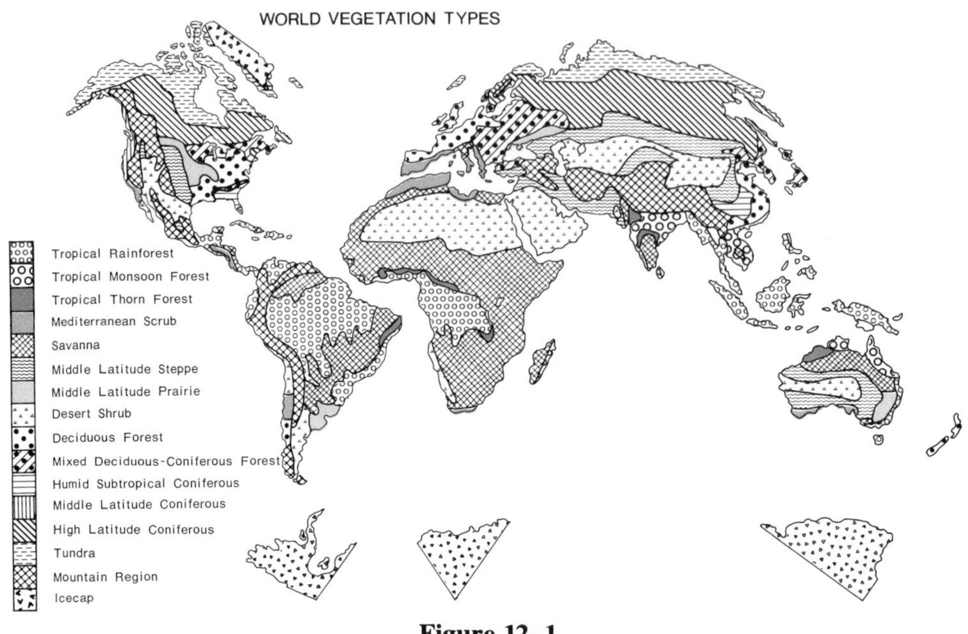

Figure 12–1

the world maps of climate and soils presented in previous modules, you should be able to see a distribution pattern very much like that on the other maps.

Knowledge of the distribution of natural vegetation around the world is important because it increases our awareness of the entire arrangement of the separate components that compose our physical environment.

OBJECTIVES

By the end of this module, you should be able to do the following:

1. List the fourteen types of climax vegetation.
2. Describe the world distribution pattern of vegetation in terms of changes of climate and/or soil.
3. Briefly list the basic climatic characteristics associated with each vegetation type.
4. Describe the major characteristics of each vegetation type.
5. List the economic advantages of each vegetation region.

KEY TERMS

selva	scrub
canopy	muskeg
broad-leaved	herbivore
lianas	carnivore
epiphyte	herbaceous
jungle	succulents
semideciduous	permafrost
sclerophyllus	

Now you are ready to begin the audiovisual portion of this module. Select the MEDIAPAK 12 component(s) and proceed. Following is a topical outline of the audiovisual sequence. You will find this outline helpful for reference and review. After completing MEDIAPAK 12, return to this book to perform the exercises.

OUTLINE

Types of climax vegetation
>*Forests*
>>Tropical rainforest (selva)
>>Tropical monsoon
>>Tropical thorn
>>Mediterranean scrub
>>Humid subtropical coniferous
>>Middle-latitude deciduous
>>Middle-latitude coniferous
>>Middle-latitude mixed coniferous and deciduous
>>High-latitude coniferous
>*Grasslands*
>>Tropical savanna
>>Middle-latitude steppe
>>Middle-latitude prairie
>*Desert shrub*
>*Tundra*

Exercises

1. The two major climatic elements that determine vegetation distribution are
   A. temperature and pressure.
   B. temperature and winds.
   C. winds and solar radiation.
   D. moisture and pressure.
   E. moisture and temperature.

2. Broad-leaved evergreen trees are features primarily of the
   A. tropical monsoon forest.
   B. middle-latitude mixed forest.
   C. subtropical coniferous forest.
   D. tropical thorn forest.
   E. selva.

3. The terms *chapparal* and *maquis* are associated with the
   A. Mediterranean forest.
   B. savanna.
   C. tropical monsoon forest.
   D. taiga.
   E. middle-latitude steppe.

4. The giant sequoia and the Douglas fir are two of the important species for lumbering in the
   A. subtropical coniferous forest.
   B. Mediterranean forest.
   C. middle-latitude mixed forest.
   D. middle-latitude coniferous forest.
   E. boreal forest.

5. Grasses become the dominant vegetation form due to
   A. fire.
   B. increased temperatures.
   C. decreased moisture.
   D. soil type.
   E. none of the above.

6. Which grass grows in clumps 4 to 8 inches high interspersed with open ground and low bush?
   A. Steppe
   B. Prairie
   C. Savanna
   D. Tundra

7. Which vegetation form is associated with permafrost?
   A. Steppe
   B. Taiga
   C. Tundra

    D. Middle-latitude mixed forest
    E. Desert shrub

8. Xerophytic features characterize certain plants in all but which of the following vegetation regions?
    A. Desert shrub
    B. Tropical monsoon forest
    C. Mediterranean scrub forest
    D. Tropical thorn forest

9. Major production of corn and wheat and raising of hogs and beef cattle are features associated with which vegetation region?
    A. Steppe grass
    B. Middle-latitude deciduous forest
    C. Subtropical coniferous forest
    D. Middle-latitude mixed forest
    E. Prairie grass

10. Highlands may affect vegetation by influencing both temperature and moisture.
    A. True
    B. False

11. Sedges, mosses, and lichens, much standing water, and few woody plants characterize the taiga regions.
    A. True
    B. False

12. The tropical rainforest includes the largest variety of plant species of any vegetation region.
    A. True
    B. False

13. Large herds of zebras, wildebeests, and antelopes, along with lions, cheetahs, and elephants, are closely associated with the
    A. jungle.
    B. tropical savanna.
    C. tropical monsoon forest.
    D. taiga.
    E. tropical rainforest.

Summary

There are fourteen types of climax vegetation which vary due primarily to differences in climates and soils. These fourteen types are listed in Table 12–1. We will discuss each of these categories in the order listed.

Table 12–1. *Types of Climax Vegetation.*

Forests
1. Tropical rainforest (selva)
2. Tropical monsoon
3. Tropical thorn
4. Mediterranean scrub
5. Humid subtropical coniferous
6. Middle-latitude deciduous
7. Middle-latitude coniferous
8. Middle-latitude mixed coniferous and deciduous
9. High-latitude coniferous

Grasslands
10. Tropical savanna
11. Middle-latitude steppe
12. Middle-latitude prairie

13. *Desert Shrub*

14. *Tundra*

Forests

Forests in the world vary in both the variety of tree species and the number of trees. Forests of moist tropical regions generally have many large trees and an outstanding array of tree species. As climatic conditions become more restrictive farther poleward, the number of tree species declines significantly.

The *tropical rainforest,* or **selva,** is the most typical forest of humid tropical climatic regions (see Figure 12–2). The forest is composed of distinct vertical zones of vegetation, capped at the top by the intertwining crowns of large trees. The largest trees can be 150 to 200 feet tall. The interwoven crowns of these large trees form a nearly unbroken **canopy** over the forest which restricts penetration of sunlight to lower levels. The canopy is a year-round feature of the forest because the trees are evergreen all year. These trees are not only evergreen but also broad-leaved; therefore, they are termed **broad-leaved evergreens.** Broad-leaved evergreens will lose their leaves during the year, but never their entire supply at one time.

Due to a continuous canopy cover which limits sunlight, the tropical rainforest floor usually has sparse vegetation growth. Sharing the canopy are large woody vines, called **lianas,** which climb the trunks of trees in search of sunlight, and many **epiphytic plants,** which anchor themselves to the trees and draw their nourishment from the air. Below the upper canopy is a second layer of vegetation made up of smaller scattered trees which grow where there

Figure 12–2

is a break in the canopy and sunlight filters through. Most of these smaller trees are shade-tolerant and rarely exceed 50 feet in height. The lowest two layers of the rainforest are composed of scattered shrubs and ground plants such as mosses. Overall, the forest floor is relatively clean of vegetative matter due to rapid decay caused partly by insect activity. **Jungle vegetation,** a mixture of entangled tall grass, low brush, bamboo, and tiny trees, grows in those few areas where sunlight can penetrate to the forest floor; jungle is particularly common bordering river courses, or where fire has destroyed the overhead canopy.

The tropical rainforests of the world contain a substantial resource of valuable tree species for lumber. However, the inaccessibility of many of these regions and the growth habits of the trees make lumbering a very expensive operation.

In the *tropical monsoon* regions, the forest has a slightly different character. The monsoon forest contains many of the same species as the rainforest, but because of the dry season, the forest does not have quite as dense a canopy and many trees are deciduous rather than evergreen. Therefore, more sunlight penetrates to the forest floor and generates dense vegetation growth at the lower levels. Indeed, jungle is much more characteristic of the monsoon forest than of the selva. During the dry season, many of the trees lose their leaves. Consequently, this forest is classified as tropical **semideciduous.** The monsoon forest is somewhat better suited for lumbering than the rainforest because trees often grow in stands (large groups) rather than as separate individuals, particularly the teak trees.

As the length of the dry season increases, the tropical forest becomes totally deciduous; this transition between the tropical monsoon forest and the tropical grasslands is the *tropical thorn forest*. The appearance of the tropical thorn forest varies from region to region due to differences in rainfall. In some areas the forest is relatively dense with moderate-sized trees, whereas in other areas the term *forest* hardly seems applicable because grass is more plentiful than the few scattered trees in the area. Most of the species of the thorn forest are adapted to drought conditions, for they have xerophytic characteristics common to many desert plants. These characteristics include small size (the trees rarely exceed 25 feet in height), thick and waxy leaves, thick bark, thorns, and deep tap roots. In the wet season the forest is green, lush, and often quite dense, but during the dry season it turns brown and appears lifeless (see Figure 12–3).

Figure 12–3

At subtropical latitudes, two forest types are found: the Mediterranean scrub forest and the humid subtropical coniferous forest. The *Mediterranean scrub forest* is directly related to Mediterrean climatic regions. In spite of the dry summers in these regions, total annual precipitation is adequate for some tree species, provided they are adapted to drought conditions. Unlike the species of the tropical thorn forest, which are also adapted to drought, most of the species of the Mediterranean forest are broad-leaved and evergreen. The trees remain evergreen because of xerophytic adaptations, such as small size, thick and leathery leaves, thick bark, and deep root systems. The trees of the forest are usually widely scattered and often possess distorted shapes. The cork oak, olive, and chestnut trees are some common examples. As shown in Figure 12–4, between the trees in the Mediterranean forest is the dominant

Figure 12–4

vegetation—**scrub** or bush. Low woody brush, called *chapparal* in California and *maquis* in Europe, is well adapted to dry summer conditions. It grows profusely in the wet season as new roots grow directly off a deep tap root. In summer it is dry and brown and presents a considerable fire hazard. Fire has been a common feature in Southern California during many recent summers.

Humid subtropical coniferous forest is located in regions of humid subtropical climate. Rainfall is more abundant in this climate than in any other subtropical region, in some cases exceeding 40 inches annually. The rainfall is generally uniformly distributed throughout the year. Subtropical coniferous forest is closely associated with sandy soils. Pines are better able to adapt to sandy soil than are broad-leaved trees, because their lateral root systems are capable of capturing incoming moisture before it infiltrates into the subsoil. Loblolly, yellow, and longleaf pines are the most common varieties. Most trees grow to heights of 50 to 100 feet with moderate density (see Figure 12–5). In the United States, these pine forests serve a thriving lumber industry, and many of the present forest areas have been planted. The byproducts of the lumbering industry, such as turpentine, pine tar, and resin, support an important secondary industry.

The forests of the middle and high latitudes include four types: middle-latitude deciduous, middle-latitude coniferous, middle-latitude mixed coniferous and deciduous, and high-latitude coniferous, or taiga. The absence of extensive regions of mesothermal and microthermal climates in the Southern Hemisphere restricts these forests largely to the Northern Hemisphere. The distribution pattern of these forests is controlled basically by the increased severity of the winter season. Subtropical forests gradually blend poleward

Figure 12–5

into broad-leaved deciduous, middle-latitude coniferous, and mixed forests. Further poleward, the mixed forest grades into high-latitude coniferous.

Middle-latitude deciduous forests grow in regions of mesothermal climate and in soils that are not too sandy (see Figure 12–6). In North America, the forest is commonly composed of an oak-hickory, oak-chestnut, or walnut-poplar association. Trees reach moderate heights of 50 to 75 feet, and smaller trees form a rather dense undergrowth. In North America, as one travels poleward into regions of humid continental climate, the forest changes to a mixed forest of broad-leaved deciduous and coniferous trees (see Figure 12–7). Birch, maple, and beech replace much of the oak-hickory association, and they are interspersed with various pines, spruce, and fir. The mixed forests of the middle latitudes have largely been destroyed by agricultural pursuits, and the forests at these latitudes are being replanted generally with fast-growing conifers.

Middle-latitude coniferous forest is a feature only of the Northern Hemisphere. It is located in association with marine west coast climate, and therefore it covers much of the Pacific Northwest coast of North America. The combination of moderate temperatures and abundant precipitation is ideal for the growth of large coniferous trees. Huge species of trees are common to this forest, including the giant sequoia or redwood, the Douglas fir, and the hemlock. As shown in Figure 12–8, many trees grow to heights in excess of 200 feet and have trunks more than 35 feet in diameter at their base. Tall ferns, some reaching 8 feet high, occupy the spaces between the large trees. The forests of the Pacific Northwest have long been the source for the largest lumber producers in North America.

Figure 12–6

Figure 12–7

Figure 12–8

Figure 12–9

High-latitude coniferous forests are referred to as *taiga* in Europe and Asia, and as *boreal forest* in North America. These coniferous forests are closely associated with subarctic climatic regions and the mean position of the 50°F isotherm. The high-latitude coniferous forest is the most extensive forest in the world. It is composed mostly of coniferous evergreens such as spruce, fir, and pine, although the larch, a coniferous-deciduous tree, grows in abundance in the USSR. The boreal forest is dense along its southern margin but thin and scattered along its poleward edge. Few trees exceed 50 feet or have trunks wider than 1 foot in diameter. The trees have pointed crowns, and thus sunlight penetrates to the lower branches where a low canopy is built over the forest floor. The forest floor is littered with twigs, branches, needles, and old trees due to the slow decay of organic materials in this climate. Interspersed with the forest are open sections of bog, called **muskeg,** or exposed bare rock, both of which are due to glacial action (see Figure 12–9).

Grasslands

The reduction in the availability of water is largely responsible for regions dominated by grasses rather than forests. The three major grasslands in the world are the tropical savannas, the middle-latitude steppes, and the prairie grass regions.

Tropical savanna vegetation is directly related to tropical savanna climate. Only a few trees can survive a six-month dry season, and therefore grass dominates the vegetation cover. During the wet season, grasses may exceed 10 feet tall and be lush and dense. With the onset of the dry season, the grass dries up into a brownish, parched, and sharply edged vegetation cover. Grass fires, which occur naturally and are also set by people, restrict the growth of bush vegetation. In Africa, savanna grasses are a vital food source for large **herbivores** such as zebras, wildebeests, antelopes, and elephants. Herbivores are kept in check by the **carnivores**—leopards, lions, cheetahs, hyenas, and others.

Middle-latitude steppe is a transitional vegetation lying between the moister prairie grass regions and the arid desert regions. The steppe grass regions are associated with steppe climates. Steppe grass forms an interrupted vegetation cover of clumps of grass 4 to 8 inches tall separated by open ground, often with low-growing shrubs also interspersed. These regions are usually too arid for agriculture and have been traditionally utilized for grazing cattle or sheep.

Bordering the steppe grass regions in more humid regions lie regions of *prairie grass*. As shown in Figure 12–11, prairie grass forms a nearly uniform cover of grasses ranging from 1 to 2 feet in length. Most prairie grass is located in humid continental climatic regions that receive 20 to 30 inches of

Figure 12–10

rainfall annually. Prairie grass regions have been prime agricultural lands since the early 1800s, when the steel plow was introduced. Corn and wheat are the most common grains harvested, with much of the corn production used to support hog and beef cattle businesses. The major drawback to agriculture in these regions is periodic drought.

Desert Shrub

The dry climatic regions of the world generally support a *desert shrub* vegetation community. Desert shrubs are adapted to lack of moisture through various xerophytic characteristics, including small size, deep tap roots, wide-

Figure 12–11

spread lateral roots, often spines instead of leaves, and soft-tissued stems or very thick bark. Deserts can be places of great color contrast, appearing brown and lifeless until it rains, at which time they bloom into a myriad of brightly colored flowers. The variety of plant life in the desert is truly remarkable (see Figure 12–12). Although most plants are small in stature, some, such as the saguaro cactus and the Joshua tree, may be 50 feet tall. Some woody plants are deciduous to conserve moisture. Other plants, called **succulents,** such as cactus, store water in their tissue.

Figure 12–12

Tundra

Poleward of the taiga and the boreal forest lie the tundra climatic regions. In this climate, trees will not grow and woody plants of any kind are rare. Vegetation in tundra regions must be able to live through only a two- to three-month growing season; extremely long, cold winters; cold, moist soils low in nitrogen; and potential frosts, even in summer. The few varieties of plants that are capable of growing under such conditions comprise *tundra vegetation*. Tundra vegetation consists of low-growing grasses, sedges, mosses, lichens, and various other **herbaceous** plants. Much of the ground surface is constantly wet or swamplike due to the **permafrost** (permanently frozen subsoil) which prevents moisture from infiltrating the soil. Despite such a harsh environment, there is a considerable variety of tundra vegetation. Desert tundra, dwarf scrub tundra, and bog organic tundra, for example, all differ in composition and appearance.

Questions

1. Why do the varieties of plant species generally decrease as latitude increases?

2. What are the vertical zones in a tropical rainforest?

3. How does jungle vegetation compare to typical selva vegetation?

4. What are the difficulties of lumbering in tropical rainforest regions?

5. How does the tropical monsoon forest differ from the selva?

6. How can you explain an evergreen forest in a Mediterranean climatic region?

7. What are the major tree species associated with the subtropical coniferous forest? Of what importance are these trees to human activities?

8. What factors make the Pacific Northwest a prime lumbering region in the world?

9. Why are middle-latitude forests features primarily of the Northern Hemisphere?

10. How do the broad-leaved deciduous species in a purely deciduous forest differ from those in a mixed forest?

11. What is the major cause of the lack of trees in the tropical savanna?

12. How does steppe grass differ from prairie grass? What is the economic importance of each of these vegetation regions?

13. What xerophytic characteristics does desert shrub vegetation have?

14. What factors restrict vegetation development in tundra regions?

15. What is the importance of the location of the 50°F isotherm to the distribution pattern of the taiga?

Suggested Readings

ANTHES, R. et al. 1975. *The atmosphere*. Columbus, OH: Charles E. Merrill.

BOUCHER, K. 1975. *Global climate*. New York: John Wiley & Sons.

GATES, D. 1972. *Man and his environment: climate*. New York: Harper & Row.

MATHER, J. 1974. *Climatology: fundamentals and applications*. New York: McGraw-Hill.

McBOYLE, G., ed. 1973. *Climate in review*. Boston: Houghton Mifflin.

OLIVER, J. 1977. *Perspectives on applied physical geography*. North Scituate, MA: Duxbury Press.

TREWARTHA, G. 1966. *The earth's problem climates*. Madison: University of Wisconsin Press.

Module
13

**Landform
Composition**

INTRODUCTION

To most people, the landforms of the earth are stable features that change little, if any, over time. The catastrophic upheaval of the earth's surface during an earthquake and the rapid collapse of a coastline are the only major changes in the earth's surface features that we commonly hear about. However, geomorphologists, or scientists who study the earth's landforms, realize that the earth's surface is undergoing constant change and that many of these changes can have an impact on us humans.

Landform changes are caused by natural processes and by human activities. Landforms are being weathered and eroded by natural processes in one

199

region, and lifted higher in other regions. On the other hand, we have bulldozed areas for transportation or construction purposes, and we have extracted minerals from the ground, often leaving gaping holes in the landscape. We have filled coastal areas with earth materials in an effort to recapture land from the sea, and we have constructed dams that alter the natural flow of rivers.

By now you know that landforms can influence climate, vegetation, and soils, by increasing moisture in some areas and decreasing it in others and also by influencing temperature, pressure, wind, and the general circulation of the atmosphere. Thus, landforms can directly influence human activities. For example, people have settled generally along coastal regions or in moist plain regions where conditions are advantageous for agriculture, fishing, transportation, or communication. On the other hand, people have generally settled sparsely in mountainous regions because of inaccessibility, poor agricultural conditions, difficulties of communication, and a generally harsh environment.

With this module, we begin our study of geomorphology. Because the earth's landforms are composed of rocks and minerals, a general knowledge of rocks and minerals is a logical first step in understanding the complex processes that affect the shape of the earth's landforms. Thus, in this module we will discuss rocks and minerals.

OBJECTIVES

By the end of this module, you should be able to do the following:

1. Differentiate between a chemical element, a mineral, and a rock.
2. Describe and explain the rock cycle by listing both the processes that create each rock class and the characteristics of each class.
3. Provide an example from each rock class and subclass.
4. Explain the possible interruptions in the rock cycle.

KEY TERMS

element	lithification
mineral	cementation
rock	compaction
magma	dessication
lava	clastic

crystallization

intrusive igneous

extrusive igneous

porphyritic

weathering

erosion

nonclastic

metamorphism

foliated

nonfoliated

rock cleavage

Now you are ready to begin the audiovisual portion of this module. Select the MEDIAPAK 13 *component(s) and proceed. Following is a topical outline of the audiovisual sequence. You will find this outline helpful for reference and review. After completing* MEDIAPAK 13, *return to this book to perform the exercises.*

OUTLINE

Elements, minerals, and rocks

Rock cycle
    Crystallization
        Igneous rocks
            Intrusive
            Extrusive
    Weathering and erosion
    Lithification (cementation, compaction, dessication)
        Sedimentary rocks
            Clastic
            Nonclastic (chemical-organic)
    Metamorphism
        Metamorphic rocks
            Foliated
            Nonfoliated

Exercises

1. There are 92 naturally occurring chemical elements, of which silicon is the most common.
    A. True
    B. False

2. Igneous rocks form from cooling magma, a process that is termed
    A. *lithification.*
    B. *metamorphism.*
    C. *cementation.*
    D. *crystallization.*
    E. *foliation.*

3. Generally, extrusive igneous rocks have finer textures than intrusive igneous rocks.
    A. True
    B. False

4. Which of the following terms is not associated with sedimentary rocks?
    A. *Dessication*
    B. *Clastic*
    C. *Lithification*
    D. *Nonfoliated*
    E. *Compaction*

5. Which of the following is a nonclastic sedimentary rock?
    A. Granite
    B. Limestone
    C. Sandstone
    D. Marble
    E. Siltstone

6. Sedimentary rocks make up roughly _________ percent of all exposed surface rocks.
    A. 95
    B. 75
    C. 50
    D. 8
    E. 1

7. Which of the metamorphic categories exhibits rock cleavage?
    A. Clastic
    B. Nonclastic
    C. Foliated
    D. Nonfoliated

8. Which of the following is an example of a foliated metamorphic rock?
    A. Gneiss
    B. Quartzite
    C. Schist

    D. Slate

    E. All of the above except B

9. It is possible for igneous rocks to completely bypass the sedimentary stage and become metamorphic.

    A. True

    B. False

10. Due to the way they develop, metamorphic rocks are always more resistant to weathering and erosion than are igneous or sedimentary rocks.

    A. True

    B. False

Summary

Elements, Minerals, and Rocks

The variation in landform composition from place to place is of interest to the geographer because the causes of variations in and the present character of landforms affect the possible uses that can be made of the land.

All substances are composed of various chemical **elements,** of which 92 occur naturally. Two elements dominate the naturally occurring elements in the earth system: oxygen accounts for 47% of all the natural elements, and silicon, 28%. Elements combine to form **minerals.** By definition, a mineral is a naturally occurring, inorganic substance with a definite chemical and physical composition. The elements that make up a mineral have united into a crystal form that distinguishes the mineral from another. The nature of the crystal is determined by the internal arrangement of the molecules of the united chemical elements. Minerals are also differentiated by their color, hardness, and luster, and the way they break apart or fracture.

Some minerals have great value as metals, such as copper, tin, lead, nickel, iron, zinc, and tungsten. Other chemical minerals, such as nitrate, phosphate, and potash, are important additions to fertilizers. More than 2000 minerals have been identified, but about 20 compose most rock material. A **rock** is, therefore, a natural combination of minerals. Granite, for example, is a combination of the minerals quartz, feldspar, hornblende, and mica (see Figure 13–1).

Rock Cycle

To simplify the discussion of rocks, we will use a cycle of rock development to explain the evolution of rock classes. Theoretically, as time progresses, rocks advance through a so-called rock cycle and change from one rock

Figure 13–1

classification to the next. In reality, there are several interruptions in the rock cycle, but despite these interruptions, the rock cycle is convenient for explaining the variations in rocks (see Figure 13–2).

Geologists classify rocks according to how they are formed. All rocks fall into one of three classes: igneous, sedimentary, or metamorphic. The rock cycle, which describes rock formation, begins with all rock in a molten or liquid state called **magma** deep within the earth's crust (see Figure 13–2). Only there are pressures high enough to generate temperatures that can melt rock. As magma flows closer to the earth's surface, pressure and temperature decline, and the magma cools and begins to crystallize. With the complete

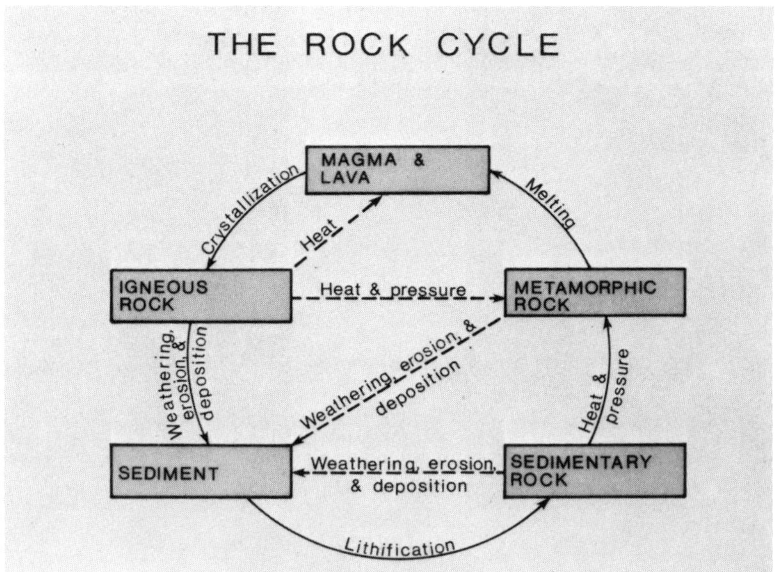

Figure 13–2

cooling of the magma, **crystallization** is complete, and solid rock has been formed. Rocks formed by the crystallization of molten rock are called *igneous* rocks. Igneous rocks may crystallize beneath the earth's surface **(intrusive igneous),** or they may crystallize on the surface exposed to the atmosphere **(extrusive igneous).**

Many geologists believe that the earth was originally a molten mass. If this is true, then all rocks were originally igneous rocks. Igneous rocks comprise roughly 95% of the outermost 10 miles of the earth's crust. Silicate minerals, a combination of silicon and oxygen plus one or more minerals, make up the bulk of magma and, in turn, compose most igneous rocks. Variations in igneous rocks are due largely to the variations in the silicate minerals that compose them and the size of crystals in the rock. Large crystals (coarse texture) in igneous rocks are the result of the slow cooling of magma and are most characteristic of intrusive igneous rocks such as granite, diorite, and gabbro (see Figure 13–3A). Small crystals (fine texture) in igneous rocks are due to the rapid cooling of lava (exposed magma) on the earth's surface and are characteristic of extrusive igneous rocks such as rhyolite, andesite, and basalt (see Figure 13–3B). Some igneous rocks crystallize partly below the surface and partly above. These rocks contain both large and small crystals and have a texture referred to as **porphyritic.** Varying mineral composition leads to many porphyritic rocks, for example, granite porphyry or diorite porphyry.

A.   B.

Figure 13–3

Igneous rocks exposed at the earth's surface are subjected to climatic elements. Temperature and moisture are capable of wearing away rocks and breaking them down in place, a process called **weathering. Erosion,** on the other hand, is the process of transporting weathered rock and organic material, which is called *sediment*. Sediment, in turn, is eventually deposited in various locations. Weathering, erosion, and sediment deposition begin the second phase of the rock cycle (see Figure 13–2).

Sediment may accumulate into very great thicknesses, and if the accumulation remains undisturbed, the lowest layers of sediment may consolidate into rock due to the extreme pressure caused by the weight of the many layers. The process of transforming sediment into rock, termed **lithification,** forms rocks that are classified as *sedimentary* rocks.

Lithification can occur through cementation, compaction, or dessication. **Cementation** is a process whereby pore spaces between sediment particles are filled with a cementing agent, which is usually a mineral dissolved in water, such as calcite, dolomite, quartz, or iron oxide. Sandstone, for example, is a rock formed from sand grains held together by a cementing agent. **Compaction** is a process common to fine-grained materials such as silt. In this process, the material itself is so finely textured that pressure is solely responsible for combining the material into a rock without any cementing agent. Siltstone is an example of a sedimentary rock formed by compaction. By the process of **dessication,** water is squeezed out of the pore spaces between fine-grained materials such as silt and clay. Water removal may be due to compaction, but evaporation of water from pore spaces is important, particularly in materials near the surface. Clay particles subjected to dessication, for example, form the sedimentary rock called *shale*.

Whereas igneous rocks are characterized by their crystal arrangements, sedimentary rocks are often distinguished by distinct strata, or horizontal layers, which are the result of the horizontal deposition of sediment in water (see Figure 13–4). Sedimentary rocks are placed into one of two broad categories: clastic or nonclastic. **Clastic** rocks are made of separate fragments of older materials such as sand, gravel, or clay. **Nonclastic** rocks, also called *chemical-organic* rocks, are derived from chemical precipitates dissolved from other rocks or from decaying organic material. A chemical precipitate is a mineral constituent that is separated from a solution by evaporation. Calcium carbonate, for example, can be dissolved from rocks by water and then collect in standing water; then evaporation occurs and the calcium carbonate and other minerals in the water consolidate into a rock. Limestone, chert, and hallite are nonclastic rocks that form from chemical precipitates.

Figure 13–4

Weathering and erosion of rock and organic material are essential to the formation of sedimentary rock. Therefore, sedimentary rocks form near or at the earth's surface. In fact, although sedimentary rocks comprise only about 8% of the rocks in the earth's crust, they comprise 75% of all exposed surface rocks.

If sedimentary rocks were to continue to be buried under increasing amounts of sediment, the bottommost layers of rock would eventually be subjected to great pressure and high temperatures. Such a change in environment could be sufficient to alter the internal characteristics of the rock. Certain minerals, for example, may melt and begin to crystallize. The altering of internal characteristics of sedimentary or igneous rocks by heat and pressure is called **metamorphism.** Metamorphism leads to the third class of rocks, the metamorphics (see Figure 13–2).

Metamorphism may alter rocks by causing chemical recombination and the subsequent growth of new minerals, by deforming the main mineral particles, or by recrystallizing minerals into larger crystals. Overall, the resulting metamorphic rock possesses a completely new appearance from the original rock.

Metamorphic rocks are classified as either **foliated** or **nonfoliated** rocks, based on the arrangement of minerals in the rock. Foliated rocks have minerals arranged in layers, so that the rock will break along roughly parallel lines. Such rocks that break along parallel lines are said to display **rock cleavage.** Slate, which is metamorphosed shale, is an example of a foliated metamorphic rock, as are phyllite and gneiss. **Nonfoliated** rocks lack any specific internal mineral arrangement, and consequently they display no cleavage. Marble, derived from limestone, and quartzite, derived from sandstone, are examples of nonfoliated metamorphics (see Figure 13–5). Examples of both foliated and nonfoliated metamorphic rocks are listed in Table 13–1.

A.

B.

Figure 13–5

Table 13–1. *Metamorphic Rocks.*

Foliated	Nonfoliated
Slate	Marble
Phyllite	Quartzite
Schist	Metaconglomerate
Gneiss	

Theoretically, if metamorphic rocks remained buried and subjected to increasing pressure and heat, they would melt completely into a molten state or, in other words, transform into magma; and the rock cycle would have gone full cycle. Rarely, however, do rocks complete the full cycle due to commonplace interruptions. A simple listing of the interruptions indicates the alternative directions in the cycle:

1. Igneous rocks may be directly metamorphosed without passing through the sedimentary stage.
2. Sedimentary rocks may be weathered and eroded into sediment and form other sedimentary rocks.
3. Metamorphic rocks may be exposed and weathered and eroded into sediment.

Questions

1. What are the two most common chemical elements in the earth system?

2. What are some of the economically important metallic minerals and chemical minerals?

3. What are the main differences between intrusive igneous and extrusive igneous rocks?

4. What percentage of the earth's crust is composed of igneous rocks? Sedimentary rocks?

5. What causes textures to vary in igneous rocks? What are the different types of textures?

6. What is the major mineral component of igneous rocks?

7. What are several examples of intrusive igneous and extrusive igneous rocks?

8. What are the various forms of lithification? Explain each process.

9. What is the difference between clastic sedimentary and nonclastic sedimentary rocks? What are some examples of each?

10. How does metamorphism alter the composition of a rock?

11. What is the difference between a foliated and a nonfoliated metamorphic rock? What are some examples of each?

12. What are the major interruptions in the rock cycles?

Suggested Readings

ADAMS, G., and WYCKOFF, J. 1972. *Landforms*. New York: Golden Press.

FOSTER, R. 1975. *Physical geology*. 2d ed. Columbus, OH: Charles E. Merrill.

KELLER, E. 1976. *Environmental geology*. Columbus, OH: Charles E. Merrill.

LEET, L., and JUDSON, S. 1958. *Physical geology*. 3d ed. Englewood Cliffs, NJ: Prentice-Hall.

STRAHLER, A. 1973. *Introduction to physical geography*. 3d ed. New York: John Wiley & Sons.

THORNBURY, W. 1966. *Principles of geomorphology*. New York: John Wiley & Sons.

VOSKUIL, W. 1969. *A geography of minerals*. Dubuque, IA: William C. Brown.

Module
14

The Shifting Continents

INTRODUCTION

For years scientists have been seeking answers to how and why the surface of the earth changes. Although their knowledge of the earth's interior is limited, they continue to explore the answers to such problems as earthquakes, the origin of continents, and the origin of the earth itself. One of the major questions has been that of movement of the earth's crust: Have continents on the earth's surface moved? If so, how and why?

The first complete hypothesis of moving continents was introduced in 1912 by a German meteorologist, Alfred Wegener. Wegener proposed that more than 200 million years ago, the continents were one supercontinent

211

called *Pangaea*. This supercontinent began to break apart in association with the rotation of the earth, and the continental pieces flowed laterally to their present positions. He cited a considerable body of evidence for continental drift, particularly the close relationships between fossils, rocks, and rock structures on opposite sides of the Atlantic Ocean. Wegener's hypothesis of continental drift created quite a stir of controversy between 1920 and 1930, and geologists refuted much of the geological evidence as coincidental. Physicists argued that rotation of the earth was not capable of moving continents and that continents could not move laterally. These scientists, however, could not offer alternative explanations for the transatlantic connections.

Wegener's hypothesis was never embraced by more than a handful of scientists during his lifetime. It was the continued research into not only geological but also climatic and botanical evidence that later generated additional support for Wegener's hypothesis. His original explanation of the mechanism for continental drift, however, was discarded in favor of the idea that movements beneath the earth's crust in the mantle layer were responsible for continental drift. Not until the 1960s, however, did the majority of the scientific community finally endorse the idea of shifting continents.

OBJECTIVES

By the end of this module, you should be able to do the following:

1. Explain how a seismograph is used to determine the interior layers of the earth.
2. Describe the interior layers of the earth.
3. Describe the two layers that comprise the earth's crust.
4. Describe the asthenosphere and the lithosphere.
5. Explain the general hypothesis that convection currents in the mantle cause plate tectonics.
6. Describe the forms of plate motion, and explain the various surface features that they can form.
7. Describe the general shift of continents that has occurred since the breakup of Pangaea.

KEY TERMS

seismograph	mid-ocean ridge
sima	subduction zone
sial	ocean trench

Mohorovičić discontinuity (Moho)   island arc

asthenosphere   fault

lithosphere   Pangaea

plate tectonics   continental shield

sea-floor spreading

Now you are ready to begin the audiovisual portion of this module. Select the MEDIAPAK 14 *component(s) and proceed. Following is a topical outline of the audiovisual sequence. You will find this outline helpful for reference and review. After completing* MEDIAPAK 14, *return to this book to perform the exercises.*

OUTLINE

Earth's interior
Inner core
Outer core
Mantle
Crust
Sima
Sial
Moho
Asthenosphere and lithosphere

Plate tectonics
Divergent motion
Convergent motion
Transform motion

History of shifting continents

Exercises

1. Which seismic wave is capable of passing only through solids?
   A. Secondary wave (S wave)
   B. Primary wave (P wave)

2. Which is the thickest layer of the interior of the earth?
   A. Inner core
   B. Outer core
   C. Mantle
   D. Crust

3. The discontinuous layer of the crust that composes the continents is the
   A. sima.
   B. sial.
   C. Moho.
   D. asthenosphere.

4. Convergent plate motion is responsible for which of the following surface features?
   A. Island arcs
   B. Sea-floor spreading
   C. Mid-ocean ridges
   D. Ocean trenches
   E. Both A and D above

5. Which of the following features is associated with convergent plate motion?
   A. Andes Mountains
   B. Japanese Islands
   C. Philippine trench
   D. Himalaya Mountains
   E. All of the above

6. Active regions of earthquakes and volcanoes are generally associated with
   A. plate interiors.
   B. plate boundaries.

7. The movement along the San Andreas fault between the North American plate and the Pacific plate is best described as
   A. divergent motion.
   B. transform motion.
   C. convergent motion.

8. Greenland separated from Europe and North American about 135 million years ago.
   A. True
   B. False

9. If plate motions continue at their present pace and direction, which statement will be true 50 million years from now?
   A. Los Angeles will be parallel to Seattle, Washington.
   B. The Pacific Ocean will continue to grow larger.

    C.   Africa will move north and cut off the Mediterranean Sea.
    D.   Australia will reconnect with Antarctica.
    E.   The Red Sea will expand all the way to the Suez Canal.

10.   Volcanic activity can be associated with both divergent and convergent plate motions.
    A.   True
    B.   False

Summary

Earth's Interior

Knowledge of the earth's interior is based primarily on **seismograph** readings that are taken to analyze earthquakes. Earthquakes are vibrations on the earth caused by the rapid release of energy produced when sections of the earth's crust shift position. From seismograph readings, scientists have speculated that the earth's interior is composed of four layers. (see Figure 14–1). At the very core of the earth is a region of extremely high-density rock called the *inner core*. The inner core is thought to be solid and possess some type of crystalline structure. Surrounding the inner core is the second layer, called the *outer core*. The outer core is assumed to be in a molten state and is likely composed of nickel and iron. The *mantle* rings the outer core and is the thickest of the interior layers. The mineral olivine makes up much of the mantle, which has both solid and molten regions. The outermost layer of the earth is the *crust,* by far the thinnest layer.

The crust can be subdivided into two sections based on composition (see Figure 14–2). Immediately around the mantle is the **sima** section, a continuous layer of basaltic rock composed mainly of silica and magnesium. The sima is exposed only in the ocean basins. The second section of the crust is a discontinuous layer that makes up the continental masses and is absent over the oceans. This layer, called the **sial,** is mostly granitic rock composed of silica and aluminum.

The thin boundary between the mantle and the crust was first identified by a Yugoslavian geophysicist, Mohorovicić. The boundary is referred to as the **Mohorovičić discontinuity,** or simply the **Moho.** Attempts to drill through the Moho into the mantle have not been successful, due largely to the difficulties of drilling into high-density rocks.

Movements of the earth's crust have been attributed, in part, to motions within the mantle. Consequently, two additional subdivisions of the interior have been defined, and they include the crust and a portion of the mantle. Between a depth of 50 miles and 450 miles is an area of the mantle referred to

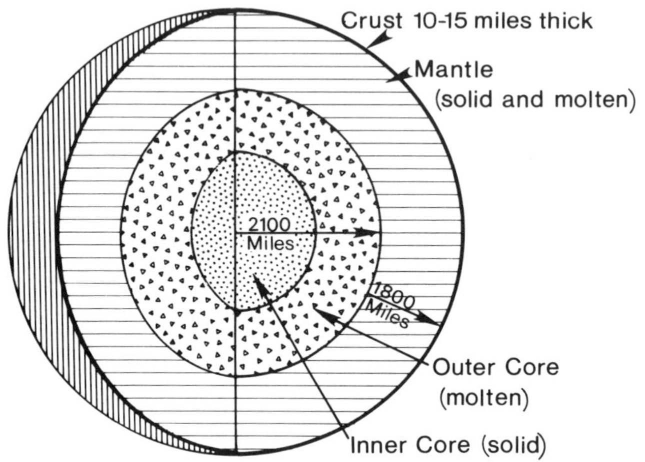

Figure 14-1

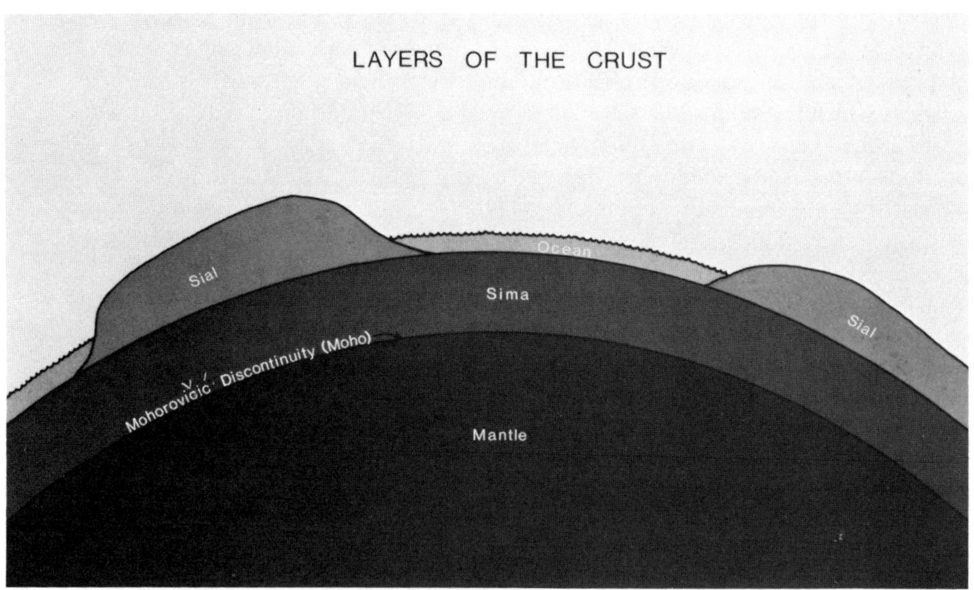

Figure 14-2

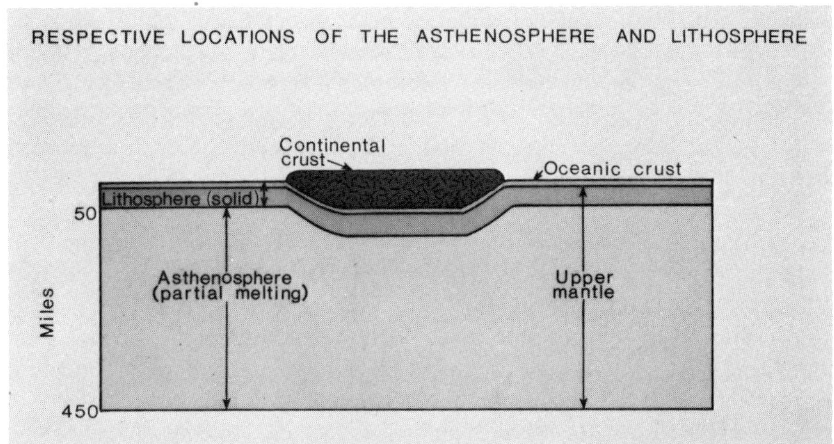

Figure 14–3

as the **asthenosphere** (see Figure 14–3). About 10% of the asthenosphere is composed of molten rock, which is a source for some of the volcanic activity on the earth's surface. Above the asthenosphere is the second subdivision, the **lithosphere.** The lithosphere is composed of solid rock, and, in a sense, it "floats" on the asthenosphere. Movements within the molten sections of the asthenosphere trigger movements in the lithosphere. Solid rock, however, does not bend and fold easily. As a result, the outer sections of the crust warp, buckle, and fracture.

Since the late 1960s, scientists have supported the concept that entire sections of the earth's crust do, in fact, shift position. Such folding, breaking, and melting of rocks within the crust is called *tectonic activity,* and the surface features produced by tectonic activity are called *tectonic features.* The lithosphere is composed of separate pieces of crust, not one large mass. These separate pieces of the crust are called *plates,* and the theory regarding plate movement is termed **plate tectonics.**

Plate Tectonics

The theory of plate movements is an extension of the earlier concepts of continental drift. The plate tectonic theory identifies nine separate plates that compose the lithosphere. Plates that include landmasses are continental plates, and those that do not are ocean plates.

Seismic readings and studies of the ocean floor suggest that the continents and the ocean floors do move and, in fact, have moved tremendous distances from their points of origin. What causes the plates to move is still not understood. The most popular hypothesis of plate movement is that convectional flow of molten material in sections of the asthenosphere results in the slow rising and sinking of the asthenosphere, which in turn triggers an adjustment of lithospheric plates; that is, the plates are forced to move up in some areas and to sink in others.

Plates shift position intact. That is, the distance between two places on the same plate remains fixed, while the distance between two places on different plates may change. All major changes, therefore, are along plate boundaries, not within the interior of an individual plate.

Three forms of plate motion are identified: divergent, convergent, and transform. *Divergent motion* occurs when two plates spread apart, creating a gap between them. Such a gap in the lithosphere allows molten rock from the asthenosphere to fill this region. The molten rock slowly cools and, in effect, forms more ocean floor. This action is referred to as **sea-floor spreading.** The Atlantic Ocean, for example, has been spreading at the rate of ½ to 3 inches per year. Where sea-floor spreading is active, there is generally a region of volcanic activity. As the gap between plates is filled by outpourings of lava from below, a ridge, called a **mid-ocean ridge,** is formed. Some mid-ocean ridges grow to elevations above sea level. For example, Iceland is an island that is part of the Mid-Atlantic Ridge (see Figure 14–4) and is a region of extensive volcanic activity.

Convergent plate motion is the second form of plate tectonics (see Figure 14–5). By convergent motion, two plates are drawn together. When they

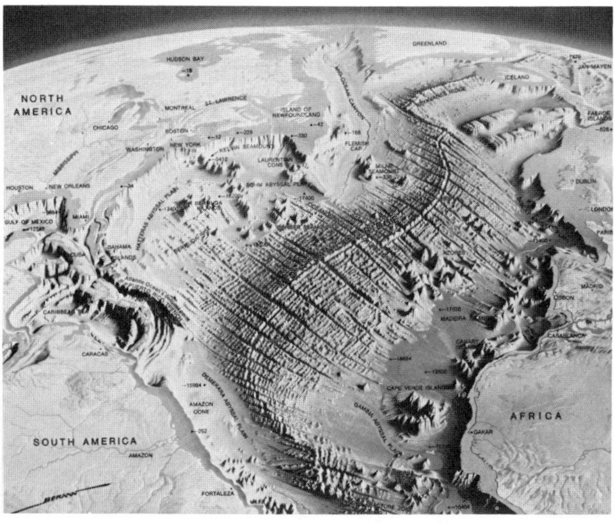

Figure 14–4

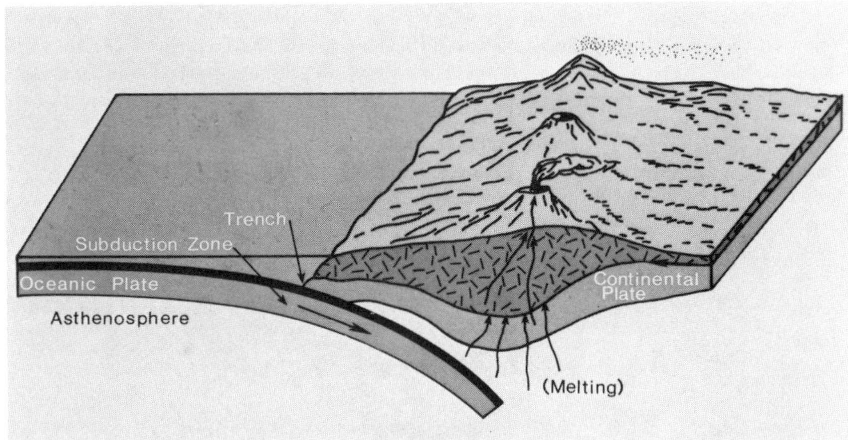

Figure 14–5

collide, generally the denser plate will sink beneath the less dense plate; this is the normal action when denser oceanic plates collide with less dense *continental* plates. As a result of this motion, the sinking plate moves into the asthenosphere, where it melts. The molten rock may find its way to the surface, and volcanic activity may result. The Andes Mountains along the west coast of South America were formed partly by volcanic activity associated with convergent plate motion. Convergent plate boundaries are also major earthquake regions.

The region where an oceanic plate sinks beneath a continental plate is referred to as a **subduction zone** (see Figure 14–5). Subduction zones may form some very deep **ocean trenches,** many of which are among the deepest spots in the ocean. For example, the Philippine trench off the east coast of the Philippine Islands is nearly 35,000 feet, or slightly more than 6½ miles, deep, and it is several hundred miles long.

When two oceanic plates converge, one plate will sink under the other plate, and this action may result in volcanic activity on the ocean floor (see Figure 14–6). If the volcanic activity continues for a prolonged period, the volcanic material may eventually accumulate to above sea level to form new land surfaces. Where a series of volcanic islands form in this fashion in the same region, the group of islands is called an **island arc.** The Japanese, the Aleutian, the Mariana, and the Tonga islands are all examples of island arcs.

The convergence of two continental plates eventually leads to subduction of one plate under the other, until there is a head-on collision of the continental landmasses (see Figure 14–7). When the landmasses ram together, the weakest rock strata buckle and may become major mountain features. In addition, there can also be volcanic activity associated with this collison as the

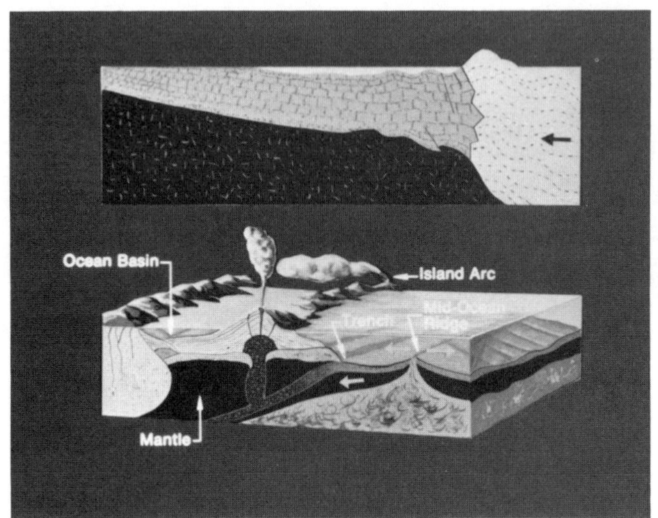

Figure 14–6

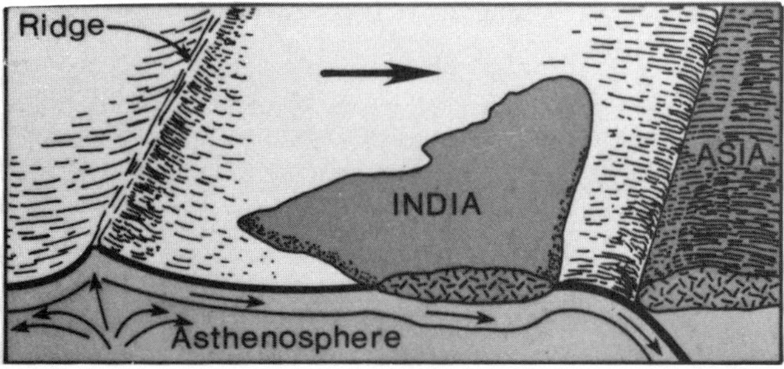

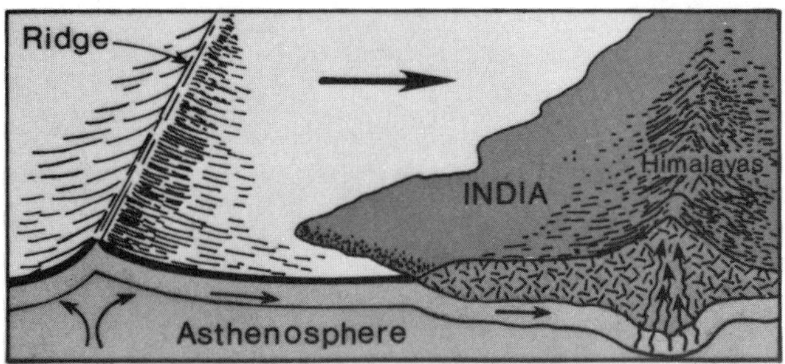

Figure 14–7

subducted plate melts and molten material rises. The collision of India with Asia is thought to have produced the Himalayas.

By the third form of plate motion, called *transform movement,* one plate slides past another along plate boundaries without any vertical motion (see Figure 14–8). A crack or joint in the rock along which such movement occurs is called a **fault.** Faults due to transform plate motion develop in the direction of plate movement. Transform faults are most common between oceanic plates, but they may be associated with two continental plates or a continental and an oceanic plate. The San Andreas fault in California is a transform fault along the boundary of the North American plate and the Pacific plate. The Pacific plate is gradually moving northwest relative to the North American plate along the San Andreas fault line.

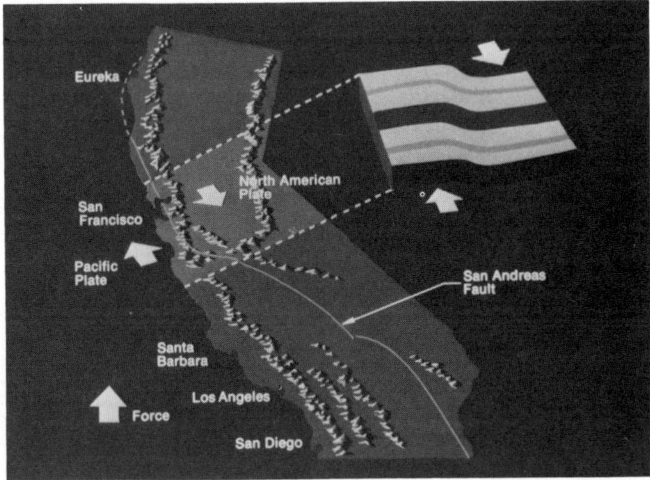

Figure 14–8

Plate Tectonics and the Continents

Most geologists now agree that originally the earth's landmasses were all connected as one supercontinent, called **Pangaea,** surrounded by one great ocean, Panthalussa, and a large sea named the Tethys Sea. Plate tectonics caused Pangaea to break up and drift, and thus the present locations of the continents are actually the result of plate motions over the past 200 million years or so.

Pangaea was located much further to the south and east than most of the present continents. For example, if New York City were in existence 200 million years ago, it would have been located at about 0° latitude and 10° east longitude. About 200 million years ago, Pangaea started breaking up into

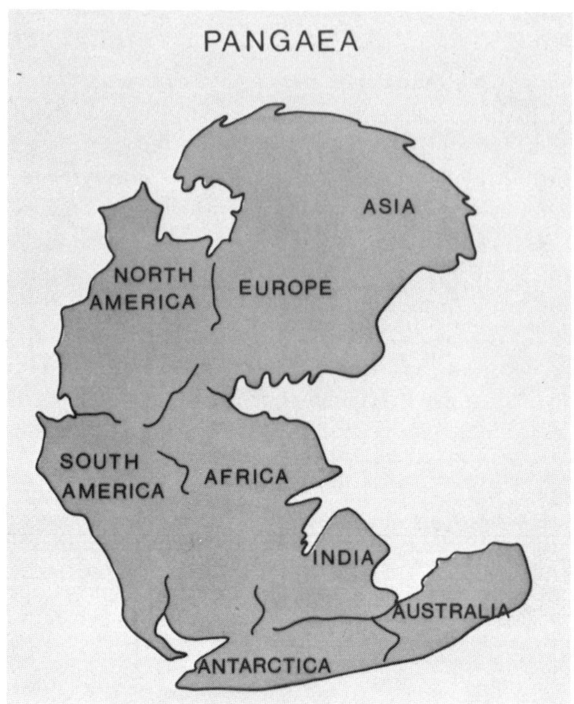

Figure 14–9

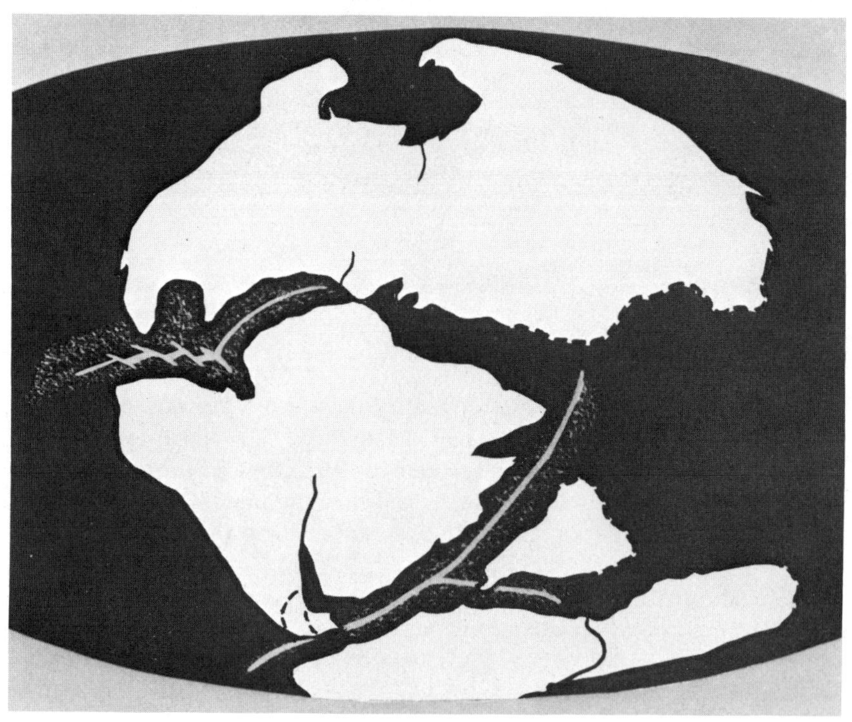

Figure 14–10

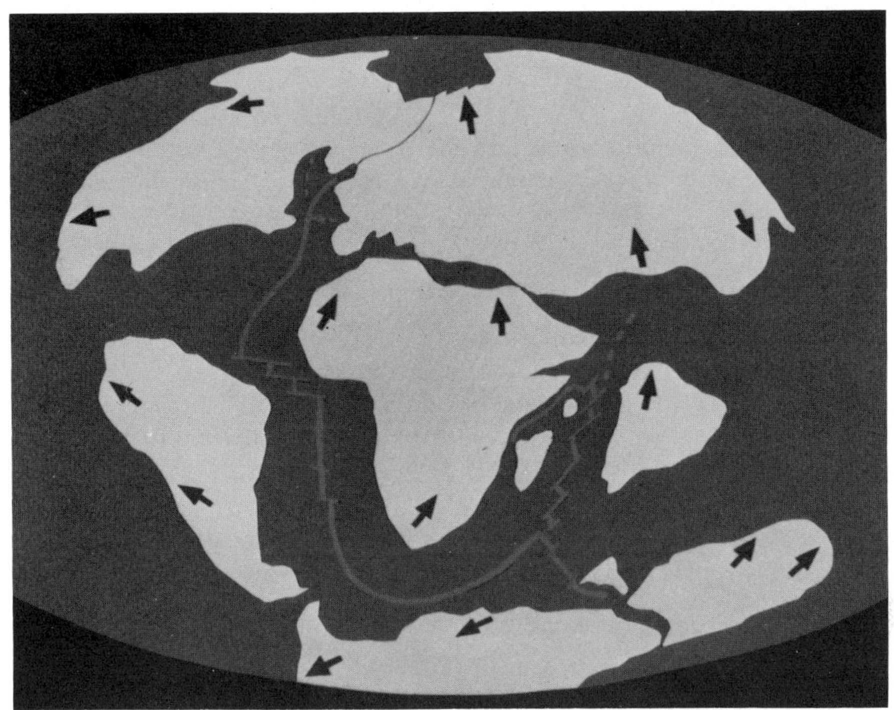

Figure 14–11

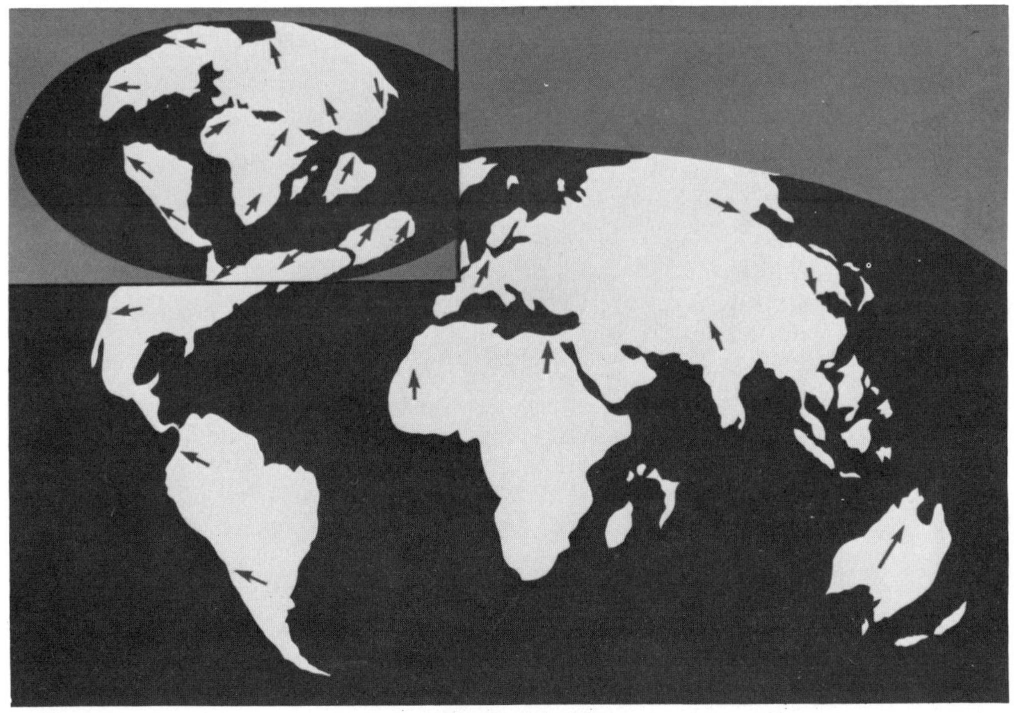

Figure 14–12

Figs. 14–11 and 14–12 from "The Breakup of Pangaea" by Robert S. Dietz and John C. Holden. Copyright © 1970 by Scientific American, Inc. All rights reserved.

separate plates. As shown in Figures 14–9 through 14–12, the continents gradually shifted toward their present locations; and, assuming that they will continue to move at present rates, we can speculate about their location 50 million years from now.

The central sections of the continents are generally areas of flat plains or plateaus because they are substantial distances from plate boundaries. These sections of basically undisturbed rock structure are called **continental shields.** Many of the shields are partially covered by sediments deposited by millions of years of erosion of upland features. In North America, the shield is called the *Canadian* or *Laurentian Shield* because the ancient igneous rocks composing the shield are exposed in many places in southern Canada. Along the margins of continental shields, sedimentary rocks are more common. It is in these regions of relatively weaker rock strata that much of the major mountain-building activity occurs.

Questions

1. How is the seismograph used to determine the nature of the earth's interior?

2. How would you describe each layer of the earth's interior in terms of mineral composition, thickness, and physical characteristics?

3. How might convection currents in the asthenosphere cause lithospheric plates to move?

4. What landform features are the result of divergent plate motion? Convergent motion?

5. How is a mid-ocean ridge formed?

6. How would you explain the formation of island arcs and ocean trenches?

7. What is the relationship between crustal plates and earthquakes? Crustal plates and volcanoes?

8. What was Pangaea? How have the continents shifted in the past 200 million years?

9. Do you think that a portion of California will fall into the sea as some have predicted? Explain.

Suggested Readings

ANDERSON, D. L. 1962. The plastic layer of the earth's mantle. *Scientific American,* July 1962. Reprint 855. San Francisco: W. H. Freeman.

————. 1971. The San Andreas fault. *Scientific American,* November 1971. Reprint 896. San Francisco: W. H. Freeman.

BULLEN, K. E. 1955. The interior of the earth. *Scientific American,* September 1955. Reprint 804. San Francisco: W. H. Freeman.

DIETZ, R., and HOLDEN, J. 1970. The breakup of Pangaea. *Scientific American,* October 1970. Reprint 892. San Francisco: W. H. Freeman.

FISHER, R., and REVELLE, R. 1955. The trenches of the Pacific. *Scientific American,* November 1955. Reprint 814. San Francisco: W. H. Freeman.

HALLAM, A. 1973. *A revolution in the earth sciences.* London: Clarendon Press.

HEIRTZLER, J. R. 1968. Sea-floor spreading. *Scientific American,* December 1968. Reprint 875. San Francisco: W. H. Freeman.

OLIVER, J. E. 1977. *Perspectives on applied physical geography.* North Scituate, MA: Duxbury Press.

TAKEUCHI, H.; UYEDA, S.; and KANAMORI, H. 1970. *Debate about the earth.* Rev. ed. San Francisco: Freeman Cooper.

TARBUCK, E., and LUTGENS, F. 1976. *Earth science.* Columbus, OH: Charles E. Merrill.

WILSON, J. T. 1963. Continental drift. *Scientific American,* April 1963. Reprint 868. San Francisco: W. H. Freeman.

Module
15

Vulcanism and Diastrophism

INTRODUCTION

During one person's lifetime, the shape of the earth's surface changes very little. Over millions of years, however, significant changes can and do occur. Several hundred million years from now, areas that are mountain regions today may be flat plain regions, and areas that are flatlands may be mountains. Also, within the relatively short, geologically, time of 50 million years, the shift of the continents will likely give the earth a completely different appearance from today.

227

All changes on the earth's surface, however, do not require millions of years. Earthquakes have disrupted the landscape by destroying cities, damming up rivers to form lakes, and disrupting transportation routes and agriculture. Volcanoes have appeared overnight in a farmer's field and have grown out of the sea in just a few days. Such changes in the earth's landforms due to alterations in the crust are results of what is called *tectonic activity*. Tectonic activity includes vulcanism and diastrophism, processes that we will define and discuss in this module.

OBJECTIVES

By the end of this module, you should be able to do the following:

1. Explain and describe the landforms that are the result of intrusive and extrusive vulcanism.
2. Explain and describe the processes of diastrophism and the landforms that they form.

KEY TERMS

vulcanism	sill
diastrophism	dike
pyroclastic	batholith
crater	stock
vent	laccolith
caldera	folding
plug	anticline
neck	syncline

aa lava   warping

pahoehoe lava   faulting

pluton   graben

concordant   horst

discordant

Now you are ready to begin the audiovisual portion of this module. Select the MEDIAPAK 15 *component(s) and proceed. Following is a topical outline of the audiovisual sequence. You will find this outline helpful for reference and review. After completing* MEDIAPAK 15, *return to this book to perform the exercises.*

OUTLINE

Tectonic activity
   Vulcanism
      Extrusive
         Volcanoes
         Lava flows
      Intrusive
         Tabular plutons
         Massive plutons
   Diastrophism
      Folding and warping
      Faulting

Exercises

1. A volcano that is wider across its base than it is high, and has gentle slopes, is called a
    - **A.** composite volcano.
    - **B.** shield volcano.
    - **C.** cinder cone.

2. The rock material ejected from an erupting volcano is called
    _________________________ material.
    - **A.** pluton
    - **B.** extrusive
    - **C.** shield
    - **D.** pyroclastic
    - **E.** composite

3. Which intrusive igneous feature is a tabular concordant pluton?
    - **A.** Dike
    - **B.** Laccolith
    - **C.** Stock
    - **D.** Batholith
    - **E.** Sill

4. When a plugged main vent of a volcano has been exposed by erosion, the exposed feature is called a
    - **A.** caldera.
    - **B.** tower.
    - **C.** neck.
    - **D.** horst.
    - **E.** pluton.

5. Which lava type forms smooth, rounded features as it cools?
    - **A.** Pahoehoe
    - **B.** aa

6. Compression forces that cause a hanging wall block to move upward relative to a footwall block form a
    - **A.** normal fault.
    - **B.** reverse fault.
    - **C.** strike-slip fault.
    - **D.** transform fault.

7. When rock strata are bent down, the resulting feature is called a(n)
    - **A.** syncline.
    - **B.** strike-slip.
    - **C.** anticline.
    - **D.** pluton.
    - **E.** caldera.

8. In which type of fault is there no vertical motion?
   A. Transform
   B. Strike-slip
   C. Oblique
   D. Normal
   E. Both A and B above

9. Grabens form
   A. hills.
   B. valleys.

10. The magma in a cinder cone has
    A. low silica content.
    B. intermediate silica content.
    C. high silica content.
    D. no silica content.

Summary

Vulcanism

Vulcanism refers to all phenomena associated with the movement of molten material within the lithosphere. Vulcanism is also often referred to as *igneous activity* because igneous rocks are formed from the crystallization of molten rock. The movement of molten material is often related to plate tectonics, and it is particularly active along plate boundaries. However, molten material may also inundate the upper layers of the crust within the central sections of crustal plates.

Vulcanism falls into two categories: *Intrusive* vulcanism is the cooling of magma beneath the earth's surface, and *extrusive* vulcanism is the action of molten rock breaking through the crust surface and cooling due to exposure to the atmosphere.

Extrusive vulcanism is of one of two basic forms: (1) the slow outpouring of lava onto the surface and (2) the rapid bursts of lava associated with volcanoes, spectacular landform features towering above the surrounding landscape. Nearly 80% of all active volcanoes are located in an uneven circle surrounding the Pacific lithospheric plate. This circle of volcanoes around the Pacific is commonly referred to as the "ring of fire." Volcanoes differ in appearance due to the composition of the molten material and the way the material is ejected onto the surface.

There are three volcano types: shield volcanoes, composite volcanoes, and cinder cones. *Shield volcanoes* are characterized by relatively nonexplosive activity due to the low silica content of the magma, which cools into basalt. Shield volcanoes grow through a series of lava outpourings. As a result, they typically have rather gradual slopes and usually are among the largest volcanoes, often extending several miles across at their base and several thousand feet high (see Figure 15–1). The five Hawaiian Islands are all shield volcanoes. Mauna Loa, overall the largest of the five, forms the island of Hawaii. It is nearly 5 miles wide at its base and 12,000 feet high.

Composite volcanoes are formed from alternating slow outpourings of lava and violent eruptions. The magma has an intermediate silica content and

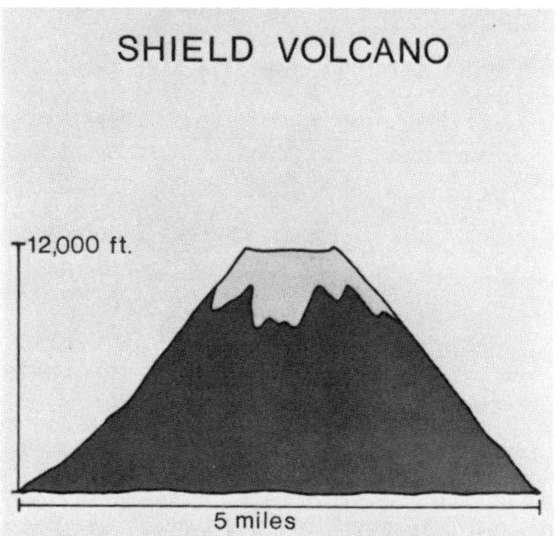

Figure 15–1

cools to form andesite. Composite volcanoes have concave slopes, sharp peaks, and, in comparison to shield volcanoes, much steeper slopes. Some composite volcanoes exceed 12,000 feet in elevation but have bases only about 2 miles across (see Figure 15–2). These volcanoes create spectacular scenery. Mt. Rainier and Mt. Shasta in the United States, Mt. Kilimanjaro in Kenya, and Mt. Fujiyama in Japan are a few examples.

Figure 15–2

Cinder cones are the most violent of the volcano forms. The lava is ejected onto the surface with such force that it is broken into tiny cinders which pile up around the central spout or **vent** of the volcano. The explosive nature of the cinder cones is attributed to the high silica content of the magma. Magma of high silica content cools into basalt or andesite. Cinder cones are usually not very large, rarely exceeding 1500 feet, and most are rapidly worn down by erosion. Often they are associated with smaller vents on the slopes of shield or composite volcanoes.

At the top and center of a volcano is the central **vent,** through which most of the lava that builds the volcano flows. When the volcano ceases to be active, the magma in the vent cools and contracts, forming a depression at the top of the volcano called the **crater.** Some craters have been enlarged by a violent eruption following a period of inactivity, or by collapse of the vent, or by a combination of both. Enlarged craters are called **calderas** (see Figure 15–3).

Figure 15–3

Erosion may strip away the outer slopes of a volcano and expose the plugged and solidified central vent. Exposed vent plugs are called **volcanic necks,** and they are a distinctive feature on the landscape (see Figure 15–4).

The lava materials deposited during violent eruptions, termed **pyroclastic** rocks, are often tiny, irregularly shaped, and loosely consolidated. Deposits of pyroclastic rocks are easily weathered and eroded due to their porous nature and their general inability to pack together. Lava flowing onto the surface in streams from cracks or fissures is called a *lava flow* or a *fissure flow.* Lava flows can be slow outpourings from tiny fissures or rapid floods from major cracks at the surface. Lava from fissure flows has covered many thousands of square miles and has even built up large plateaus: the Deccan Plateau in India is a classic example. The silica content of the lava may dictate whether the lava flow will solidify into smooth-featured **pahoehoe lava** (low silica) or jagged-edged **aa lava** (high silica).

Intrusive vulcanism is less spectacular than volcanic eruptions or even lava flows, but the landform features formed by it cover a greater total area than those formed by extrusive activity. Geologists are still attempting to explain why molten material is present in some places in the crust and not in other places. Regardless of the cause, there are known reservoirs of molten rock in various parts of the lithosphere. Pressure beneath the surface can cause magma to flow, even onto the surface. More often, however, the magma fills cracks and pockets beneath the surface.

Figure 15–4

Magma that solidifies below the surface is called *plutonic rock*. **Plutons** are classified by their size and shape, and by their relationship to surrounding rocks. Plutons that lie with their boundaries parallel to the surrounding rock strata are said to be **concordant** (see Figure 15–5); plutons that cut across surrounding rock strata are **discordant.** The shape of a pluton is either tabular or massive. Tabular plutons have thicknesses that are small relative to their length, while massive plutons are bulky in shape; that is, they are nearly as wide as they are long. The size of plutons may vary from less than an inch to several hundred feet. No separate distinction is made for size except in the massive plutons.

Intrusive features are outgrowths from a major reservoir of magma. The largest of these reservoirs may partially solidify into a discordant, massive pluton called a **batholith.** Batholiths tend to get progressively larger with depth, at least to a point. They are rarely exposed at the surface but can form the core of major mountain areas. The Coast Range of Alaska and British Columbia, for example, has a batholith core that extends 1100 miles in length and up to 100 miles in width. A **stock** is a second major reservoir of magma similar to a batholith, except that a batholith exceeds 40 square miles in size while a stock does not.

Magma flows up through cracks in rocks and fills the available spaces. Since most cracks in rocks are narrow and elongated, tabular-shaped plutons are formed (see Figure 15–5). Where the tabular pluton is concordant, the feature is called a **sill.** Sills may be only a few inches across, or they may be several hundred feet thick and extend several miles in length. Most sills are basaltic since they tend to cool rapidly due to their small volumes and large surface areas. Tabular plutons that are discordant are called **dikes.** Dikes vary from several inches to several feet across and have been known to extend 60 miles, although they usually extend 1 to 6 miles. Like sills, dikes are formed of fine-textured igneous rocks due to rapid cooling.

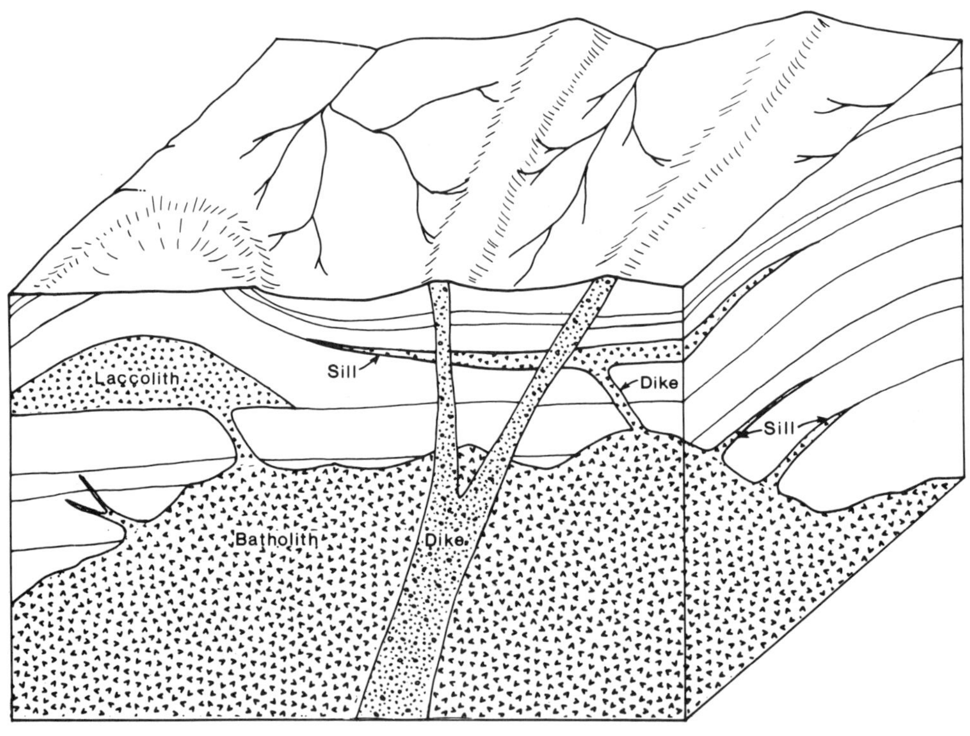

Figure 15–5

Laccoliths are another intrusive igneous feature. They are concordant, massive plutons that often appear as blisters along a sill (see Figure 15–5). They are usually composed of granite since they cool slowly due to moderate volumes and limited surface area.

Diastrophism

Diastrophism, the second type of tectonic activity, is the process or processes by which the earth's crust is deformed. The features produced by these processes are called *diastrophic* features. Diastrophism usually includes folding and warping, and faulting. The diastrophic movements of the crust are in response to pressures applied to rock due to plate tectonics or motion within the mantle.

Although it may be difficult to comprehend, hard rock can bend, much like modeling clay, without crumbling apart or without being in a molten state. You can see bent rock where noncrystalline sedimentary rocks no longer lie horizontally; the absence of crystal development indicates that the rocks were not molten. Also, you can find in some places shells and fossils of sea animals high on the tops of mountains, which indicates that the rock has been lifted to that elevation by some mechanism.

The rate at which pressure is applied to rock will determine the nature of the rock deformation. If pressure is applied in a lateral direction at a very slow rate over many hundreds of thousands or millions of years, the rocks may be compressed into a series of folds. Diastrophism of this form is called **warping** or **folding.** Folding is most common in sedimentary rocks that are relatively free of major joints or cracks. Folded rock strata that are bent up are termed **anticlines;** rock strata folded into a downward bend are called **synclines** (see Figure 15–6). Folds may be several feet across or they may form large mountains. Folding action may be so extensive that one fold completely overlaps another; such a fold is said to be *overturned.* On the other hand, the action may be so slight and gradual that the rocks are only moderately bent and may cover large areas; this type of folding is called *warping.* The Appalachians, the Alps, the Pyrenees, the Himalayas, and portions of the Rockies are all basically folded mountain regions.

The second major form of diastrophism is **faulting.** Faults are joints or cracks in rock along which movement has occurred. The joints are produced as magma cools, or they are formed gradually by variations of temperature and moisture. Most joints are near to the earth's surface, and they become fewer and smaller with increased depth. Faulting develops when the lateral pressure on rock strata is applied rapidly and unevenly. Rocks that do not

Figure 15–6

bend easily will give along zones of weakness, namely, the joints. Thus, instead of folding or bending, the rocks move up, down, or laterally along a fault. Faulting may displace rocks only a foot or two, but with time, it can move rocks several thousand feet and build mountain ranges.

Movements along faults are described relative to the two rock layers on either side of the fault line (see Figure 15–7). The rock layer above the fault line is termed the *hanging wall block,* while that below the fault is the *footwall block.* The motion of these two rock layers may be either up or down, depending on how the pressure is applied to the rocks. If the hanging wall block moves down in relation to the footwall block, the fault is termed a *normal* or *gravity* fault. Normal faults are the result of tensional stress produced by stretching of the crust in relation to vertical lifting. The term *normal* is a misnomer because most faulting is not a result of tensional forces. Normal faults generally produce a low fault scarp, or slope, that usually is rapidly destroyed by erosion. Hence, these features are difficult to identify on the surface.

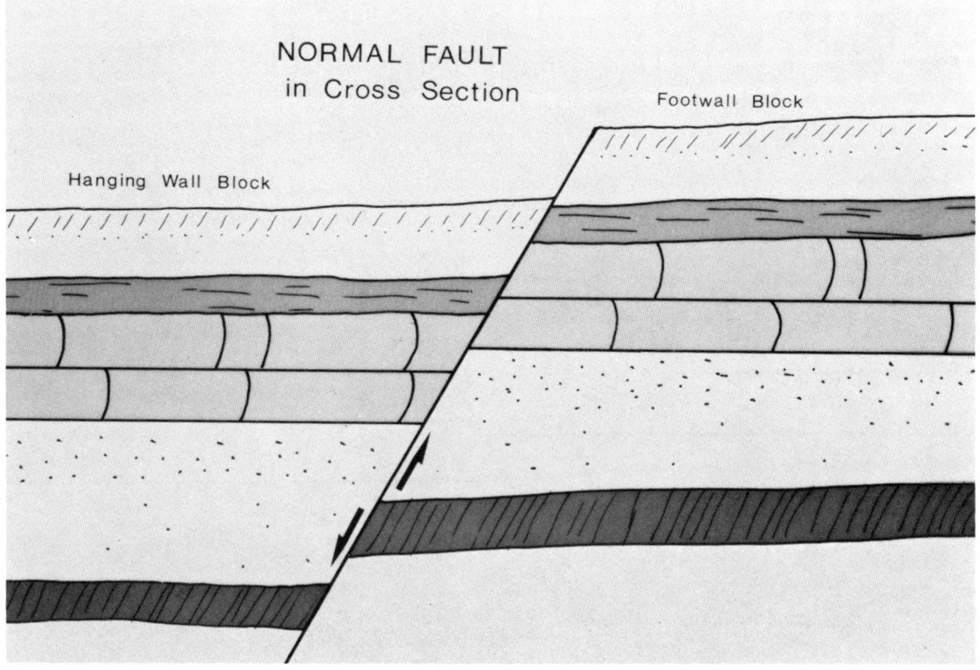

Figure 15–7

Reverse, or compression, faults are the result of compressional forces that push rocks together (see Figure 15–8). As a result of compression, the hanging wall block moves up along the fault line above the footwall block. At the surface, a slope or fault scarp is built which can vary from several feet to many thousands of feet in height. One type of reverse fault, called a *thrust* fault, results when a fault line is at a very low angle so that the hanging wall block often slides completely on top of the footwall block.

Faults that display lateral motion are called *strike-slip* or *lateral* faults. (see Figure 15–9). Extreme lateral faulting is called *transform* faulting. *Oblique* faults display both lateral and vertical movements.

Active tensional forces may cause blocks of rock to sink between two parallel fault lines. The result is a long, narrow valley called a **graben.** The Rhine Valley, and the Rift Valley of East Africa, are two examples of a graben. The raised blocks on either side of a graben are called **horsts** (see Figure 15–10). Horsts may result when compression pushes a block up between parallel faults.

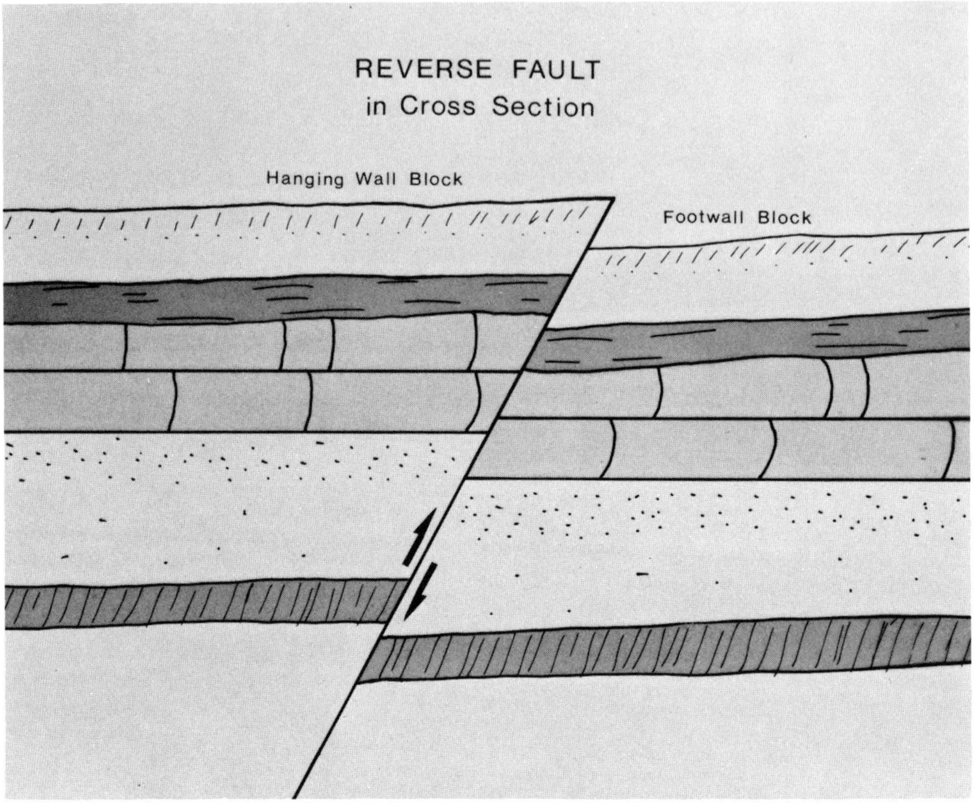

Figure 15–8

Figure 15–9

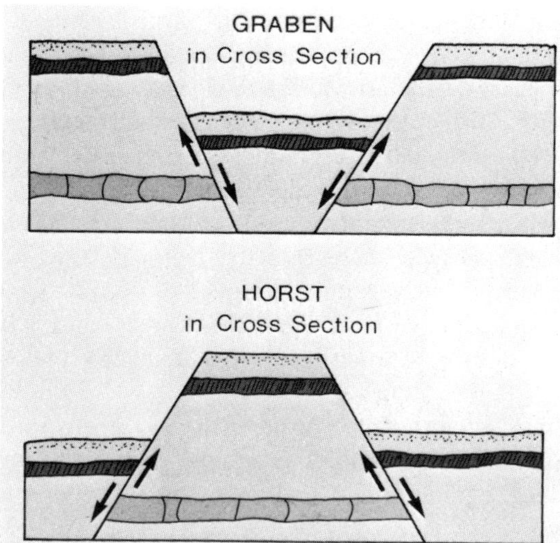

Figure 15–10

Faulting over long periods of time can build mountains called *fault-block* mountains. The Sierra Nevadas of California, parts of the Andes of South America, the Atlas Mountains in northwestern Africa, and portions of the Rockies are just a few examples.

Folding and faulting, as well as vulcanism, commonly occur together in the same region. Therefore, many areas have complicated geologic structures due to the mixture of diastrophism and vulcanism.

Additional Material

Earthquakes

Today the most dramatic evidence of movements in the earth's crust is the sudden displacement of rock along fault lines, movement that results in vibrations of the crust called an earthquake. An earthquake does not cause displacement; it is the shock waves set in motion by rock displacement that are the earthquake. Fortunately, most earthquakes are so slight that you would have trouble even noticing them. Other earthquakes, however, may rattle dishes, cause buildings to collapse, or initiate landslides and avalanches.

An earthquake radiates out in all directions from its center, or focus, in the form of shock waves. The energy of the earthquake rapidly dissipates with increased distance from the focus. The sudden displacement of rocks along a fault is the release of tremendous pressure that has built up in the rocks for many years. Upon the release of pressure during displacement, the rocks snap back to their original shape, much like the bending of the shaft of a hunting bow: With the release of the arrow, the bow shaft snaps back to its original position. The action of rocks returning to their original shape is called *elastic rebound*. It is elastic rebound that produces earthquake vibrations.

The shock waves of an earthquake are in the form of surface waves, S waves, or P waves. S waves and P waves are important to the study of the earth's interior, but surface waves are the major destructive waves of an earthquake. As surface waves travel through the ground, they cause the ground to move up and down much like ocean waves cause a boat to rise and fall. It is the strength of the surface waves that determines the potential destructiveness of the earthquake.

The magnitude of earthquake shock waves can be measured by the Richter scale. This scale is based on the total energy released by an earthquake as measured by a seismograph. The magnitudes on the Richter scale vary from 1 to 10. For every 1-unit increase in magnitude, ten times greater total energy is released. The largest earthquakes have magnitudes near 8.6, while those less than 2.5 are usually not detected by humans. The Alaskan earthquake in 1964 had a magnitude of 8.3, as did the San Francisco earthquake in 1906. The Peruvian earthquake of 1970 had a magnitude of 7.8 and resulted in the deaths of more than 50,000 people. Major earthquakes in Nicaragua in 1972, and in Turkey in 1975, both had magnitudes greater than 7.5, and in both cases many people died and the destruction of homes and other buildings was extremely heavy.

Figure 15–11

Scientists hope to be able to predict earthquakes someday. In several instances, they have been able to identify regions of great earthquake potential. Much more knowledge is required, however, before scientists will be able to guarantee to people that an earthquake will strike the area in which they live within the next day or two, the next month, or even the next year.

Questions

1. What are the three types of volcanoes? How do they differ from one another?

2. Why do some lava flows form smooth, rounded features and others rough, jagged features?

3. What is the "ring of fire"?

4. Can you list several volcanoes and areas of major lava flows in the world?

5. What is the relationship between vulcanism and plate tectonics?

6. What are the various features formed by intrusive vulcanism? How do these features become exposed at the surface?

7. How do folding and faulting differ?

8. What kinds of landforms are the result of folding? Faulting?

9. What are the various types of faulting? Describe how each forms.

10. How are grabens and horsts formed?

11. What are some major folded mountains and fault-block mountains?

Suggested Readings

BARNEA, J. 1972. Geothermal power. *Scientific American,* vol. 226, no. 1.

BULLARD, F. M. 1962. *Volcanoes: in history, in theory, in eruption.* Austin: University of Texas Press.

EATON, J. P., and MURATA, K. J. 1960. How volcanoes grow. *Science,* October 7, 1960.

MACDONALD, G. A. 1972. *Volcanoes.* Englewood Cliffs, NJ: Prentice-Hall.

MILNE, L. J. et al. 1962. *The mountains.* Life Nature Library. New York: Time Inc.

OLIVER, J. E. 1977. *Perspectives on applied physical geography.* North Scituate, MA: Duxbury Press.

WILLIAMS, R. S., and MOORE, J. G. 1973. Iceland chills a lava flow. *Geotimes* 18.

Weathering and Mass Movement

INTRODUCTION

Many of the variations in the earth's landforms are due to the gradual wearing away of rocks and soil exposed to the elements of weather. Both rocks and soils wear away at differential rates because of differences in their internal composition and because of the variable nature of the environment around the world.

The process of wearing away the earth's landforms is called *degradation*. Degradation, which also includes the removal of the materials worn away by the elements of weather, gradually lowers the elevation of the land. Degradation is accomplished by the agents of running water, moving ice, wind, and gravity. In this module and the next few, you will learn how each agent operates to modify the earth's landforms.

OBJECTIVES

By the end of this module, you should be able to do the following:

1. Explain the differences between weathering, erosion, and deposition.
2. Describe and explain the various types of mechanical weathering.
3. Describe and explain the various types of chemical weathering.
4. Describe and explain the types of landslides.
5. Describe and explain the types of slow mass movement.

KEY TERMS

weathering	mudflow
erosion	earthflow
deposition	slump
exfoliation	avalanche
oxidation	rockfall
carbonation	talus
hydrolysis	creep
landslide	solifluction
rockslide	

Now you are ready to begin the audiovisual portion of this module. Select the MEDIAPAK 16 component(s) and proceed. Following is a topical outline of the audiovisual sequence. You will find this outline helpful for reference and review. After completing MEDIAPAK 16, return to this book to perform the exercises.

OUTLINE

Weathering
> *Mechanical weathering*
>> Temperature
>> Frost action
>> Exfoliation
>> Plants and animals
>
> *Chemical weathering*
>> Oxidation
>> Carbonation
>> Hydrolysis

Mass movement
> *Landslides*
>> Rockslide
>> Mudflow
>> Earthflow
>> Avalanche
>> Slump
>> Rockfall
>
> *Slow mass movement*
>> Creep
>> Solifluction

Exercises

1. The breakdown of rocks in place into smaller pieces of the same rock best describes
   A. erosion.
   B. mechanical weathering.
   C. chemical weathering.

2. Which type of weathering is most effective in regions of warm, humid climate?
   A. Mechanical weathering
   B. Chemical weathering

3. Which type of weathering is in fact the result of both mechanical and chemical weathering?
   A. Hydrolysis
   B. Frost action
   C. Oxidation
   D. Carbonation
   E. Exfoliation

4. Which rock type is particularly susceptible to carbonization?
   A. Granite
   B. Marble
   C. Basalt
   D. Limestone
   E. Shale

5. The initial motion of earth materials in mass movement is due to
   A. water.
   B. gravity.
   C. wind.
   D. moving ice.
   E. weathering.

6. Which is not a type of rapid mass movement?
   A. Mudflow
   B. Earthflow
   C. Solifluction
   D. Slump
   E. Avalanche

7. Which type of mass movement involves the shift of a block or blocks of material along a curved plane of flow?
   A. Slump
   B. Creep
   C. Earthflow
   D. Rockslide
   E. Solifluction

8. Tilted telephone poles, gravestones, and fences on gentle slopes are the few types of evidence of
   A. slump.
   B. mudflow.
   C. creep.
   D. rockslide.
   E. earthflow.

9. In which location would solifluction most likely be active?
   A. Rocky Mountains
   B. Great Plains
   C. Florida Everglades
   D. New England coastline
   E. Alaskan tundra

10. Landslides, over long periods of time, move much more earth material than do slow mass movements.
    A. True
    B. False

Summary

Weathering

The earth's landforms are constantly subjected to the elements of weather. As a result, rocks are broken down and worn away, and landforms undergo steady modifications. The breakdown of rock and soil in place is called **weathering,** and the action of removing the weathered materials is called **erosion. Deposition** is the laying down of the materials transported by erosion. There are several agents of erosion: gravity, water, ice, and wind. Mass movement is the erosion and deposition caused by the force of gravity.

Weathering is completed by both mechanical and chemical actions. *Mechanical weathering,* also called *disintegration,* is the breakdown of rocks and soil into smaller pieces. Mechanical weathering alters *only the size* of the weathered material; that is, a granite boulder will be broken down into granite pebbles. Mechanical weathering, however, can assist chemical weathering by increasing the number of exposed surfaces on which chemical action can work.

There are several types of mechanical weathering. The expansion and contraction of materials due to *temperature change* can result in the breaking apart of rock and soil materials. The continual stretching and shrinking of rocks help weaken the chemical bond between the minerals in the rock. Thus, areas where temperatures fluctuate greatly on either a daily or a seasonal basis are prime locations for this type of weathering. Deserts and mountain regions are good examples, for both commonly experience large daily temperature ranges.

Frost action, along with temperature variation, is a major type of mechanical weathering. Water collects in the joints of rocks, and as it freezes, it expands in volume by 9%, pushing out against the sides of the rock joint. This outward pressure may be sufficient to gradually break the rocks apart. More likely, however, rocks are broken apart by a combination of freezing-thawing and expansion-contraction activities over extended periods of time.

Exfoliation is a process whereby rocks are broken into eggshell-like pieces; that is, the breaking away of one layer exposes a new spheroidal surface (see Figure 16–1A). Exfoliation is the result of expansion and contraction, as well as the slow chemical breakdown of minerals in rocks by water. It most commonly affects igneous rocks, such as granite, diorite, and gabbro, and massive rocks, such as marble. Large-scale exfoliation forms large, rounded mountains called *exfoliation domes,* many of which are extremely prominent features, such as Half Dome in Yosemite National Park (see Figure 16–1B), Stone Mountain in Georgia, and Sugar Loaf in the Rio de Janeiro harbor.

Plant roots and *burrowing animals* can also trigger mechanical weathering. Tree roots grow into rock joints and help pry the rocks apart. Burrowing animals can dig beneath a rock mass, weaken the rock's foundation, and cause the rock to collapse under its own weight. Also, decaying organisms often release acids that can assist chemical weathering.

Chemical weathering, unlike mechanical weathering, breaks rock apart by destroying its internal composition; as a result, the rock is altered from its original form. Chemical weathering is accomplished largely in association with water. Water combines with other substances, such as oxygen or carbon dioxide, and forms mild acids that are capable of dissolving certain minerals. One type of chemical weathering is called **oxidation,** a process whereby oxygen, dissolved in water, is combined chemically with various minerals. Certainly you are familiar with the oxidation of iron into iron oxide, or rust. Rocks containing iron are often stained yellow or brown as the iron oxide is dissolved from the rock mass. As a result, the rock mass is weakened and may begin to fall apart.

Carbonation is a type of chemical weathering by carbonic acid, which forms when water combines with carbon dioxide. Carbon dioxide is available in the atmosphere, where it can be absorbed by raindrops, and it is also released by decaying organic matter. Limestone is especially susceptible to carbonation because the mineral calcite, which makes up much of limestone, readily dissolves into calcium carbonate, or lime, and is rapidly dissolved from limestone. Limestone is particularly susceptible to carbonation

Figure 16–1

in humid regions, where it generally appears deeply pitted and grooved.

Hydrolysis, a third type of chemical weathering, is the permanent chemical union of water and a mineral. Igneous rocks, for example, are very susceptible to this type of breakdown. The mineral feldspar, common in granite, combines into a clay mineral when combined with water; since clay minerals are very soft, the rock structure is weakened.

Chemically weathered rocks assume a new appearance due to the alteration of their internal components. Since most chemical weathering involves water, the general result is that rocks swell slightly as they are weathered. This action, called *spheroidal weathering,* tends to produce rounded features (see Figure 16–2).

Figure 16–2

The rate at which rocks weather depends upon the internal characteristics of the rock and the nature of the environment in which the rock is located. Some rocks and minerals are more resistant to environmental changes of temperature and moisture than are others, and, of course, the environment can vary considerably from place to place. Mechanical weathering is the primary weathering form in regions where temperatures are moderate to cold and moisture is seasonal or generally sparse. Chemical weathering is most active in regions of consistently high temperature and abundant moisture. The differential rate of weathering of rocks helps to explain the variation in the earth's landscape. Resistant rocks often form uplands beside lowlands that were formed as weathering stripped away the rocks at a faster rate than in the upland areas.

Mass Movement

Mass movement, also called *mass wasting,* is the movement of earth material downhill due to the pull of gravity. Once rock has been weathered, the pieces of rock that have broken away are capable of moving down any available slope. The quantity and speed of material transported by mass movement vary with the type of earth material and the degree of slope. In some cases, whole mountainsides have given way and roared downhill in just a matter of moments. More often, earth materials move grain by grain, only fractions of an inch at a time. Even so, over long periods of time, considerable quantities of earth materials are transported downhill.

Gravity is responsible for initiating mass movement, but once materials are in motion, their flow can be assisted by water. Water reduces the surface tension between particles, thereby weakening the bond between particles. In addition, water will also add weight to a rock or soil mass, which by itself may be sufficient to overcome the surface tension in the material.

As steepness of slope increases, so also does the probability that mass movement will occur; in addition, as steepness of slope increases, so also does the speed of mass movements. Mass movements are particularly common in new construction site areas, for example, where bulldozers can steepen the existing slopes while simultaneously removing the vegetation cover which assists in holding soil and rock in place on a slope.

Mass movements are classified as either rapid or slow. Rapid mass movements, called **landslides,** may bring great destruction and cause loss of life. Slow mass movements, on the other hand, operate so slowly that they cannot be detected by the human eye.

Landslides are of varying types: rockslides, mudflows, earthflows, slumps, avalanches, and rockfalls. **Rockslides,** the most potentially destructive of all mass movements, are the rapid movement of rock strata. Often weak layers of clay, shale, or limestone lie parallel to a slope or have joints and fractures parallel to the slope; the layers collapse and no longer support the rock layers above them, thus causing the rocks to break away from the rock mass and move downslope. A rockslide can be triggered by an earthquake, by gradual weathering of rock layers, or sometimes by just the added weight of a new building. Rockslides involving entire portions of a mountainside have been known to give way and wipe away all forms of life in their paths.

Mudflows are another form of landslide that can be very dangerous. A mudflow is a mixture of rock, soil, and water that flows downslope. Although most mudflows do not result in great destruction, there have been several that have been particularly devastating to sections of homes. The hills surrounding Los Angeles have experienced quite a number of mudflows ever since people, in order to construct their homes, stripped away the sparse vegetation holding the earth materials in place. **Earthflows** are similar to mudflows. They slide over a bedrock surface as they move and are particularly common on steeper slopes within humid climatic regions. They may involve only a few cubic yards of material or occasionally as much as several million cubic yards (see Figure 16–3).

Slumping is the downward and outward movement of rock and debris traveling as a unit or a series of units. By the action of slumping, large blocks

Figure 16–3

of material give way at the same time and travel along a curved plane. The upper surface of the block tilts backward as it moves. Water often assists slump by accumulating along the curved plane of flow. Slumps normally involve only several cubic yards of material but, like other landslides, may also involve several thousand cubic yards of debris (see Figure 16–4).

Figure 16–4

Landslides composed of ice and snow are called **avalanches.** Confined to high mountain regions with steep slopes and abundant snow, avalanches are, because of their location, of little threat to human life. Nevertheless, they can be very powerful and are certainly capable of destroying or burying most objects along their paths. The National Park Service expends considerable effort each year in triggering potential avalanches with dynamite to protect skiers.

The last type of landslide, called **rockfall,** is the tumbling of individual pieces of rock downhill. These pieces accumulate at the base of a slope and can build a considerable pile of rock debris called talus. The talus may be sufficient to form a secondary slope called a *talus slope.* Talus slopes are common in most mountain regions.

Whereas all forms of landslides can be easily detected by people, slow mass movements are often more difficult to identify. There are two types of slow mass movement: creep and solifluction. **Creep** is the movement of soil and weathered rock down gentle slopes (see Figure 16–5). There are no scars or breaks in the surface materials of slopes to suggest that creep has occurred or is occurring. There is, however, evidence of creep in the appearance of trees and man-made objects such as retaining walls, telephone poles, fences, and gravestones. Gravestones and fences may be tilted on slopes, while trees may have a slight curve to the base of their trunks. It generally takes several years before the evidence of creep can be detected.

Figure 16–5

Solifluction is a feature of the higher latitudes where the subsoil is permanently frozen, a condition called *permafrost*. In the summer in regions where permafrost is common, water saturates the upper layers of the soil because the permafrost prevents meltwater and rain from percolating downward. Under saturated conditions, only a slight slope is necessary to trigger the movement of the earth materials. The terrain of tundra regions is especially modified by solifluction.

Slow mass movements are certainly less spectacular than landslides, but, in the long run, they are capable of moving as much material as, if not more than, rapid mass movements do.

Questions

1. How do mechanical and chemical weathering differ?

2. How do expansion and contraction help to weather rock? What other weathering process is closely associated with expansion and contraction?

3. What are the types of chemical weathering? Give an example of each.

4. In what environments do chemical and mechanical weathering predominate?

5. What is meant by *differential rates* of weathering?

6. What factors assist mass movement?

7. What factors can trigger a rockslide? What might be the cause of a rockslide?

8. How can people affect the incidence of mudflows and earthflows?

9. How does creep affect the landscape?

10. What conditions are necessary for solifluction to be active?

Suggested Readings

FOSTER, R. J. 1975. *Physical geology*. 2d ed. Columbus, OH: Charles E. Merrill.

LEET, L. D., and JUDSON, S. 1958. *Physical geology*. 2d ed. Englewood Cliffs, NJ: Prentice-Hall.

OLIVER, J. E. 1977. *Perspectives on physical geography*. North Scituate, MA: Duxbury Press.

STRAHLER, R. N. 1973. *Introduction to physical geography*. 3d ed. New York: John Wiley & Sons.

THORNBURY, W. D. 1954. *Principles of geomorphology*. New York: John Wiley & Sons.

Landforms from Running Water

INTRODUCTION

Running water is, overall, the most effective landform modifier. It continually carves away uplands and streams, and transports eroded materials eventually to the oceans. Stream erosion causes major changes in the size and shape of landform features in almost all climatic regions. Even in desert climates, brief thunderstorms in a few days can erode more material than do winds throughout an entire year. Each year, billions of tons of stream-collected materials are carried to the oceans, and coastal landforms are modified as stream-transported materials are deposited in large quantities where the streams meet the sea. Were it not for tectonic activity that creates new uplands, the earth would long ago have been eroded into a flat, featureless plain, lying at sea level.

We depend on running water for many things: for recreation, energy, fresh water, and waste disposal. Unfortunately, our use of the world's streams and rivers has generally resulted in the deterioration in the quality and quantity of water. Many rivers are no more than moving sewers, and polluted rivers flowing into the sea are having an increasingly detrimental effect on the ecological balance of ocean life. Dams have been constructed across rivers without proper long-range planning as to their effect on plant and animal life, and even as to their benefits for people. The Aswan Dam is a good example.

Construction of the Aswan High Dam, built across the Nile River in Egypt in the 1960s, initiated a well-publicized controversy regarding a prized archeological site, which would have been made inaccessible under several hundred feet of water in the proposed Lake Nasser behind the dam. The public furor was so intense that at considerable expense to the government, the tombs were painstakingly dissected, moved to a new site, and reassembled. However, as it turned out, archeology was not the only problem created by the dam. For thousands of years, the flood waters of the Nile had annually deposited large quantities of fertile alluvium along the course of the river, and people had lived in rhythm with the river's changes. Denied the replenishment of alluvium from the annual floods, the farmers below the dam found their soils rapidly depleted of fertility. Crop yields began to drop significantly, and the government began subsidizing farmers with extremely expensive artificial fertilizers. In addition, the enormous quantities of alluvium now collect behind the dam in such high volume that the estimated life expectancy of the dam has already been cut in half. Other problems due to the lack of annual flood waters include soil salinization and increased disease.

Our use of running water is not totally a negative picture, however. By modifying channels and building dams and levees, we have been able to control some of the earth's running water, thereby eliminating potentially dangerous floods and supplying fresh water for agricultural and domestic uses. But we must increase our efforts to improve the quality and quantity of running water if we expect to survive.

OBJECTIVES

By the end of this module, you should be able to do the following:

1. Explain the factors that can cause changes in stream velocity.
2. Explain the relationship between stream velocity and stream discharge.

3. Explain and describe the various methods by which a stream erodes its channel.
4. Explain and describe the methods by which stream materials are transported.
5. Explain the relationship between stream velocity and stream deposition.
6. Explain and describe the various landforms that are the result of stream deposition.
7. Explain the concept of base level.
8. Explain how a stream adjusts to either an increase or a decrease in its base level.
9. Describe the characteristics of youthful and mature stream channels.
10. Explain the development of a floodplain.
11. Describe the landforms that are the result of stream channel development.
12. Explain and describe the stages of the geomorphic cycle in humid regions.

KEY TERMS

rivulet	downcutting
stream velocity	lateral erosion
stream volume	undercut slope
stream gradient	slip-off slope
stream load	meander
stream capacity	floodplain
stream discharge	meander scar
stream mouth	oxbow lake
direct lifting	natural levee
abrasion (impact)	back swamp
cavitation	tributary
solution	yazoo tributary
suspension	distributary
bed load	rejuvenation
stream competency	terrace
saltation	denudation

(Continued)

<table>
<tr><td>alluvium</td><td>geomorphic cycle</td></tr>
<tr><td>delta</td><td>interfluve</td></tr>
<tr><td>alluvial fan</td><td>peneplain</td></tr>
<tr><td>base level</td><td></td></tr>
</table>

Now you are ready to begin the audiovisual portion of this module. Select the MEDIAPAK 17 component(s) and proceed. Following is a topical outline of the audiovisual sequence. You will find this outline helpful for reference and review. After completing MEDIAPAK 17, return to this book to perform the exercises.

OUTLINE

Measures of stream flow
Stream velocity and stream volume
  Stream gradient
  Shape and size of channel
  Relationship to materials
Stream discharge

Stream erosion
Direct lifting
Abrasion (impact)
Cavitation

Stream transport
Solution
Suspension
Bed load

Stream deposition

Base level

Stages of stream development
Characteristics of youthful and mature streams

Geomorphic cycle in humid regions

Exercises

1. Solution, suspension, and bed load are three methods of stream
   - **A.** velocity.
   - **B.** discharge.
   - **C.** transport.
   - **D.** impact.
   - **E.** deposition.

2. Generally, stream discharge decreases as channel width, channel depth, and stream velocity increase.
   - **A.** True
   - **B.** False

3. Direct lifting, abrasion, and cavitation are methods of stream
   - **A.** velocity.
   - **B.** transport.
   - **C.** erosion.
   - **D.** deposition.
   - **E.** gradient.

4. If the elevation of a stream's head is 900 feet and that of the stream's mouth is 300 feet, and the stream flows a distance of 30 miles, its average gradient is
   - **A.** 20 mi/hr.
   - **B.** 2 ft/mi.
   - **C.** 20 ft/mi.
   - **D.** 300 ft/sec.
   - **E.** 600 ft.

5. If the base level of a stream is lowered, the stream will most likely
   - **A.** continue to erode.
   - **B.** deposit material.

6. On which side of a meander in a stream channel will erosion be most active?
   - **A.** Slip-off slope
   - **B.** Undercut bank

7. Lateral cutting and the formation of a floodplain are characteristic of a(n)
   - **A.** youthful stream.
   - **B.** mature stream.
   - **C.** old-age stream.

8. Yazoo tributaries are most closely associated with
   - **A.** alluvial fans.
   - **B.** oxbow lakes.
   - **C.** interfluves.
   - **D.** natural levees.
   - **E.** deltas.

9. An abandoned meander that contains water is called a(n)
   - **A.** meander scar.
   - **B.** distributary.

    C. oxbow lake.
    D. interfluve.
    E. terrace.

10. Alluvial fans are depositional features of mountainous regions.
    A. True
    B. False

Summary

Erosion by water, or stream erosion, modifies existing landforms by stripping away weathered materials. In addition, streams form as they deposit eroded materials. Stream erosion is capable of transporting debris, eroding more deeply the channels in which the stream flows, and depositing large volumes of sediment.

Measures of Stream Flow

A stream's abilities to transport, erode, or deposit are largely functions of its velocity and volume. **Stream velocity** is the distance the water travels in a unit of time; it is usually measured in meters per second or feet per second. **Stream volume** is the amount of water in the stream channel. Both velocity and volume vary from stream to stream and within the same stream at different times.

Stream velocity is affected by **stream gradient,** the rate of change in elevation from the stream's source to its mouth. It is usually measured in feet per mile or meters per kilometer. It generally decreases downstream toward the **mouth** of the stream. The larger the gradient, the greater is the stream velocity. Portions of some rivers have gradients less than 1 foot per mile, while others, such as mountain streams, may have gradients of several hundred feet per mile.

Channel shape and size also affect stream velocity. Since the flow of water is opposed by friction between the water and the sides and bottom of the stream channel, the less the water comes into contact with the walls and bed of the stream, the less will be the friction and the greater will be the velocity. Semicircular channels are best suited for higher velocities because friction is least (see Figure 17–1).

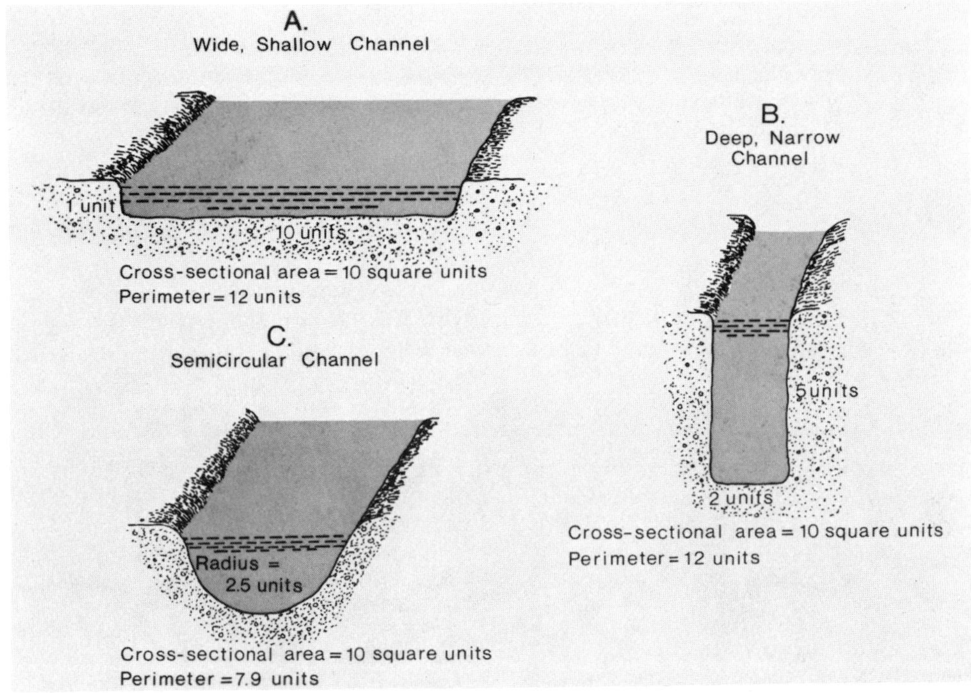

Figure 17–1

Stream load, or the materials carried by the stream, can also affect velocity because of its relationship to friction. Stream loads can vary from fine clays and silts to gravel, rocks, and even boulders. The larger the solid particles transported by a stream, the rougher is the channel and the greater is the amount of friction.

Stream velocity is one indication of stream flow. Closely related to stream velocity, and also a measure of stream flow, is stream discharge. **Stream discharge** is the amount of water flowing past a given point in a given unit of time; it is usually expressed in cubic feet per second. Stream discharge rates vary from high values during periods of flooding to low values when the amount of water is limited. At times of high discharge, stream channels become wider and deeper to handle the increase in volume, which in turn improves the flow of water by reducing friction and therefore increasing velocity. During periods of low water, discharge and velocity are reduced.

Stream velocity and discharge vary not only with seasonal changes but also from place to place along the stream's course. Discharge usually increases downstream, which means that generally stream channels increase in size downstream to handle the increase in water. In addition, on the average, velocity increases downstream because of the increased efficiency of larger channels and a corresponding decrease in friction between the stream load and the stream channel.

Stream Erosion

Streams are highly effective erosion agents; they remove material from the banks and the bed of their channel in several ways. When water flows in a turbulent fashion, the overturning water can dislodge particles from the chan-

Figure 17–2

nel. This form of stream erosion is called **direct lifting.** Large particles cannot be directly lifted without high stream velocities. Sand particles can be directly lifted at velocities of ¼ mile per hour (1 foot per second), while pebbles can be directly lifted when velocities reach 1 mile per hour (10 feet per second). Very fine particles, such as clay, may require high velocities in order to be picked up, because they pack together so well and do not extend much into the channel.

Particles transported in a stream act as tools of erosion as they collide with the stream channel and slowly chip away other particles. This form of stream erosion is called either **abrasion** or **impact,** depending on the size of the material involved. Impact involves particles larger than small pebbles, while abrasion functions with smaller particles. Impact can gouge small holes in a stream bed called *potholes*. Potholes are features of streams with waterfalls and rapids where velocities are high and the flow of water is turbulent. Rocks and stones in such streams are usually well rounded because abrasion has worn away any rough edges (see Figure 17–2).

Stream erosion is also accomplished by **cavitation,** the collapse of air bubbles in turbulent water upon collision with the stream channel. Cavitation is active only when stream velocity exceeds about 30 feet per second. The burst of air bubbles against the stream channel is like little dynamite blasts. Indeed, some experiments have suggested that this force is as great as 2000 pounds per square inch. Like erosion by impact, cavitation is most active in streams with rapids and waterfalls where the water motion is turbulent.

Stream erosion, whether it be by direct lifting, abrasion, impact, or cavitation, is most effective during flood stage, when velocity and discharge are at a maximum.

Stream Transport

Once material has been eroded by a stream or supplied to the stream by mass movement, the material is transported downstream. The amount of material that the stream is carrying at any given moment is the **stream load.** The maximum amount of material that a stream can transport is called *capacity;* few streams maintain a capacity load. Transported materials are carried in **solution,** by **suspension,** or by **bed load,** depending on the size of the material and the stream velocity. The largest material that a stream transports is a measure of the stream's **competency.** The relationship between stream competency and velocity is a multiple relationship. If stream velocity is doubled, for example, the stream's competency increases four times. If velocity is tripled, competency increases nine times, and so forth.

Transportation of materials in **solution** involves particles that are dissolved in water, such as various minerals and organic matter. Dissolved materials are measured in parts per million. The quantity of material transported in solution varies with the climate, the composition of the stream channel, and the additions to the stream by groundwater. Solution may appear to transport an insignificant amount of material, but it is estimated that nearly 300 million tons of material are transported in this fashion in the United States annually, and about 3 billion tons of dissolved material flow to the oceans from around the world each year.

Material transported in **suspension** generally includes fine sand, silt, and clay particles, although larger particles can be suspended during flood stages (see Figure 17–3). About two-thirds of the material transported by streams is in suspension. The particles remain suspended in the stream as long as velocity is adequate. As velocity decreases, the stream's ability to suspend materials decreases, and the size of the suspended material proportionately decreases.

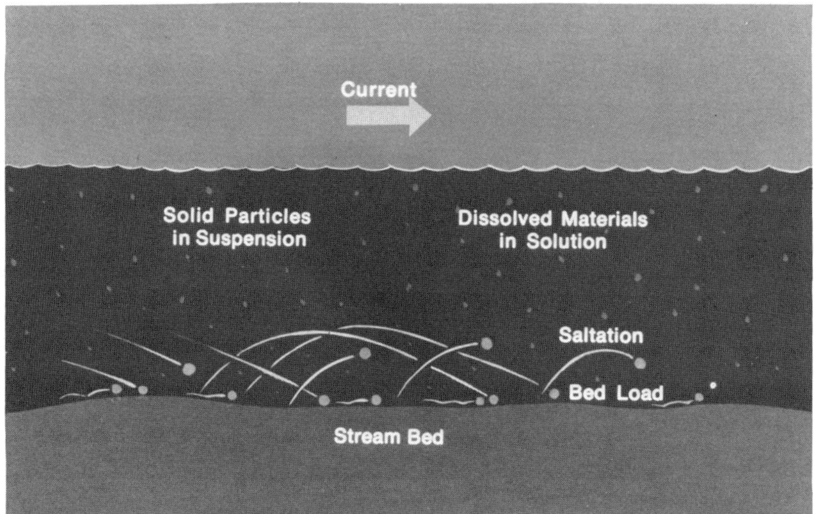

Figure 17–3

Bed load is the material pushed by moving water along the stream bottom; it is composed of particles that are too heavy to be suspended. A stream can transport bed load by sliding or rolling it along the stream bottom, or by using another method called **saltation.** Saltation is the action whereby a particle is temporarily lifted free from the stream bed, carried in suspension a few inches or feet, and then dropped back to the stream bed. The action is repeated many times as the particle hops its way downstream. About 10% of stream material is transported as bed load, although during flood stage as much as 50% can be transported in this manner.

A reduction in stream velocity results in a decrease in the transporting ability of a stream. When stream velocity decreases, a part of the stream load is deposited at the initial point of reduced velocity, with the heaviest materials being the first to be deposited (see Figure 17–4). If stream velocity continues to decline, progressively smaller particles are deposited, until eventually the stream enters standing water, such as a lake or the ocean, and the finest suspended particles gradually settle out onto the lake or ocean bottom. Such stream-deposited material is called **alluvium** or **alluvial material.**

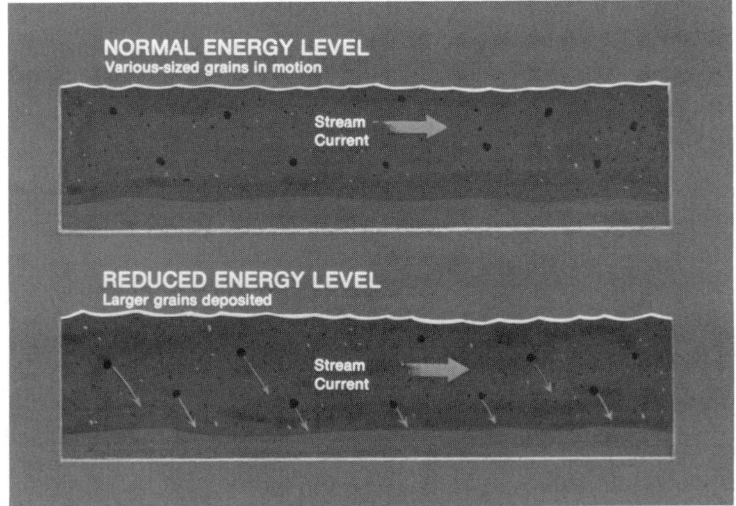

Figure 17–4

At the **mouth** of a stream or river, where the flowing stream enters standing water, the velocity of the stream is sharply reduced. The stream mouth is, therefore, an area where deposition is common. The load in the stream often spreads out into a fan-shaped deposit at the stream mouth; this fan-shaped feature is called a **delta.** Deltas are composed of alluvium, with the largest materials deposited closest to the point where the mouth begins. Deltas vary both in shape and especially in size. Some large deltas, such as the Mississippi River delta, cover about 9000 square miles (23,000 km^2) (see Figure 17–5). In dryland areas, mouths of mountain streams are at the foot of the mountains, where they often empty onto flat plains. Due to sharply reduced velocity, fan-shaped deposits, formed in the same fashion as deltas, accumulate at the mouths of these streams. Such features are called **alluvial fans.**

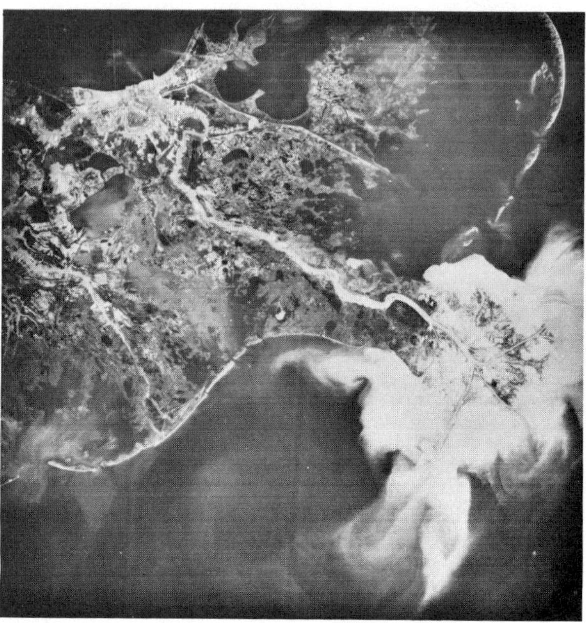

Figure 17–5

Base Level

All streams have the capacity to erode. The lowest level to which a stream can erode its channel is called its **base level.** Any factor that prevents a stream from further eroding its channel creates a base level. As a stream enters a lake, for example, its velocity quickly diminishes, and the stream loses its ability to erode. Consequently, the level of the lake becomes the base level of that stream. The lowest level to which any stream can erode is sea level. The ocean, therefore, is considered ultimate base level, even though there are a few land areas that lie below sea level. All base levels above sea level are *temporary* base levels.

The base level of a stream can change, and if it does change, the stream will adjust accordingly (see Figure 17–6). If the base level is raised, the stream gradient will be reduced; stream velocity will be reduced; deposition will increase; and gradually a newly formed stream channel will evolve. A lowering of the base level, on the other hand, will increase stream gradient, stream velocity, and the stream's ability to erode its channel. The stream channel will eventually adjust to the lower base level by changing form through increased erosion.

Reduction of Gradient

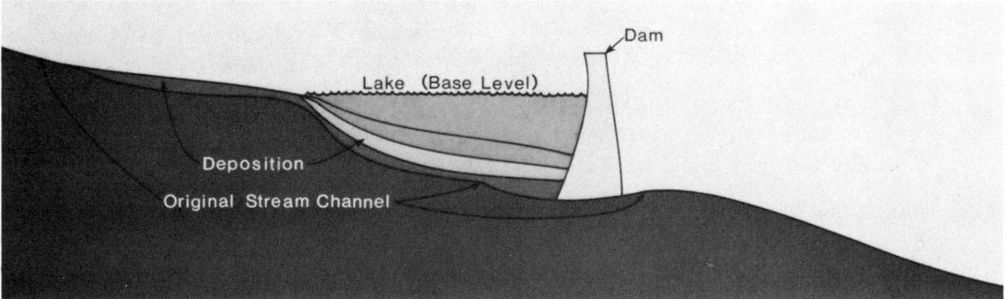

Gradient Steepened by Lowering Base Level

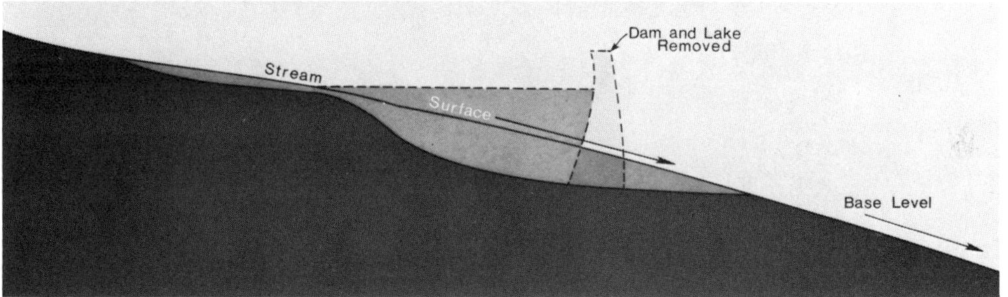

Figure 17–6

Once a stream channel has adjusted to a change in base level, it may reach a temporary condition of equilibrium in which stream discharge is balanced by stream gradient, stream width, and stream velocity. At that time, if the entire stream system is considered, stream erosion and deposition are equal. In other words, the stream simply transports the load supplied to it. It is questionable whether a stream ever attains a state of equilibrium since changes along the stream channel that can disrupt equilibrium are constantly active.

Stages of Stream Development

Stream systems evolve through a series of growth stages from youth to old age. At each stage of growth, streams display distinct characteristics not in evidence during other stages. It is important to point out before describing each stage that different portions of the same system may be in different stages simultaneously.

Streams, or portions of streams, are in the *youthful* stage when they display high velocity and are underloaded. The principal activity along a youthful stream is deepening of the channel by erosion. At this stage, the stream's ability to erode exceeds the amount of load supplied to it. Youthful streams have a distinct V-shaped profile, particularly where they cut through easily eroded rock material (see Figure 17–7). In hard rock, or alternating layers of weak and resistant rock, steep, vertical-walled gorges and canyons are youthful features.

Figure 17–7

As a youthful stream rapidly **downcuts** the channel bed, it often encounters resistant rock layers that cannot be eroded as rapidly as less resistant rock layers farther downstream; these resistant layers form ridges over which the stream plunges. Stream channels characterized by water that falls through a series of vertical steps of varying heights are called *rapids*. A waterfall is a rapid of extended vertical distance. Waterfalls and rapids are features of youthful streams, and youthful streams are most apparent in mountainous regions where stream gradients are large and runoff is abundant.

A stream evolves to the *mature* stage when its primary activity changes from vertical erosion to **lateral erosion** and deposition along the stream banks. Stream channels may reach a temporary equilibrium condition, at which time the rate at which material is supplied to the stream is equal to the average rate at which the stream can transport the material. The profile of a mature stream valley is characterized by gradually sloping sides and a nearly flat bottom in which the stream flows. The flat-bottomed stream valley is referred to as the **floodplain** because at times of high water, the river can overflow its normal banks and spread out across the valley bottom.

Floodplains expand as the mature stage progresses. Once a stream begins to flow laterally in the stream channel, centrifugal force shifts the water toward the outside of any bend in the channel. Thus, most of the energy within the stream is also directed toward the outside of bends, and it is in that portion of the stream channel, therefore, that erosion is most active. This section of the stream channel is termed the **undercut slope** (see Figure 17–8).

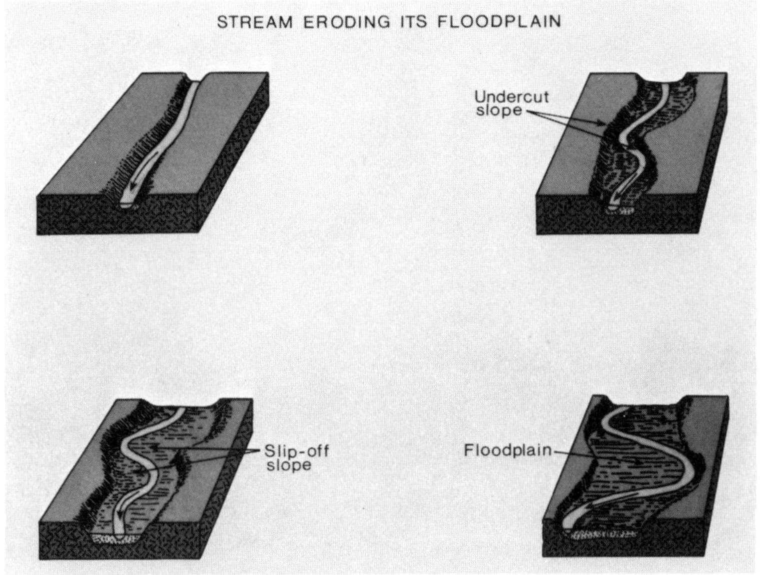

Figure 17–8

On the inside of the bends in the stream channel, energy in the stream decreases. As a result, not only is the stream's ability to erode lost, but also its ability to transport its load is reduced. It is along the inside of bends in the stream channel, therefore, that alluvium is deposited. The inside portion of the bend formed by deposition is termed the **slip-off slope** (see Figure 17–8). Uniform, sinuous curves in a stream channel are called **meanders.** A mature stream meanders back and forth as it flows downstream, undercutting one bank and depositing alluvium on the opposite bank. The floodplain is gradually expanded as slip-off slopes join together along the same stream bank.

In the middle stages of maturity, the stream load is generally high, and deposition is usually very active along the stream system. Meanders may have wound so far laterally that they can eventually turn back on themselves (see Figure 17–9). At the extreme, two meanders will join together as the stream establishes a new channel and abandons the former channel. Abandoned

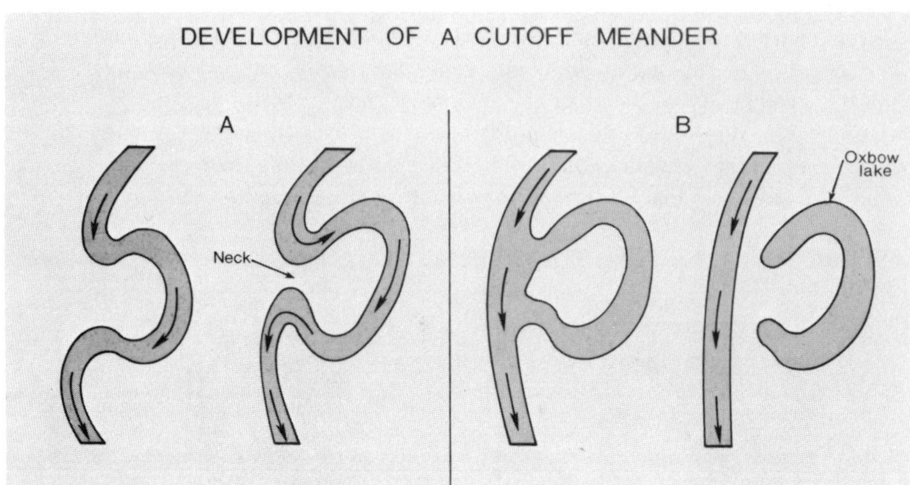

Figure 17–9

meander channels are called **meander scars.** If an abandoned meander channel continues to hold water, it forms a horseshoe-shaped lake called an **oxbow lake.** These lakes are usually spring fed.

Along a mature stream channel within the floodplain, moderately high banks called **natural levees** can form. As floodwaters recede and a stream returns to its normal size, the load that the floodwaters carry is, in part, deposited along the banks of the normal channel. The resulting natural levee, built up by alluvium, is a low, gently sloping ridge along the river course. After many years of flooding, natural levees may build up to 20 feet high or more.

The areas of the floodplain behind the natural levees are usually poorly drained. The natural levee may prevent small streams flowing across the floodplain toward the main stream channel from reaching the main channel. Consequently, the water from these streams stays impounded behind the levees and often creates swampland and marsh called **back swamp.**

Streams that feed into larger streams are called **tributaries.** Some small tributaries may flow parallel to the main channel for considerable distances behind a natural levee before they can break through and join the main channel. Tributary streams that flow distances behind levees are called **yazoo tributaries,** a name derived from the Yazoo River which flows nearly 200 miles (320 km) behind levees before it joins the Mississippi River.

Another feature of some mature streams is the breakup of the main channel into numerous smaller channels. This breakup occurs when large quantities of alluvium deposited in the main channel block the normal flow of water. Naturally, the primary area where the main channel is split up by

deposited alluvium is the stream mouth. In some large mature rivers, however, stream gradients may be so low in some sections of the stream course
that alluvium may be deposited and cause the channel to temporarily split up
into several smaller channels. Smaller streams that have split off from the
larger main stream and do not rejoin it are called **distributaries.** Distributaries
are common to deltas, and to rivers in semiarid and arid regions. The overall
stream pattern formed by numerous distributaries is referred to as a *braided
stream pattern*.

The evolution of a stream channel from one stage to the next is highly
variable. Tectonic activity is constantly at work and can cause base-level
changes that initiate adjustment in stream channels. If the landscape is lifted
by some form of tectonic activity, the landscape is said to be **rejuvenated.** A
mature stream channel, for example, may be rejuvenated and return to the
youthful stage once again. A mature meandering stream channel may renew
downcutting activity if the landscape is uplifted. A meandering river that
downcuts into bedrock due to rejuvenation is termed an *entrenched* meander
(see Figure 17–10). Adjustments in stream channels may also be triggered by
climatic changes that affect the amount of water flowing in the stream or by
glaciation, which can completely disrupt the former drainage pattern.

Mature stream channels that have had periodic reductions in volume over
long periods of time often display stream terraces (see Figure 17–11). As the
volume of a stream is reduced, the size of the floodplain that it cuts is also

Figure 17–10

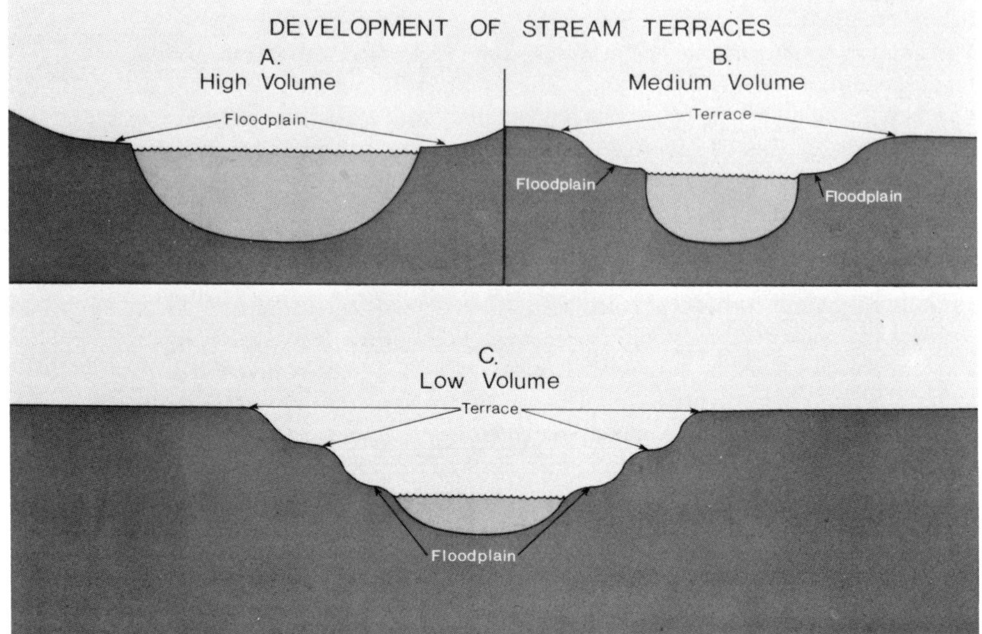

Figure 17–11

reduced. As a result, the former floodplain of the river (when it had a higher volume) stands as a flat plain above and adjacent to the new floodplain. Flat plains bordering a river and formed by adjustments in stream volume are called **terraces.**

Geomorphic Cycle in Humid Regions

Not only stream channels but all landform features theoretically evolve through a cycle of development. One theoretical cycle, called the **geomorphic cycle,** was first proposed by William Morris Davis, a famous geographer and geologist. Davis assumed that tectonic forces would not interrupt the stages of development once the initial uplift of the land surface occurred. Although this assumption is not a reflection of the real world, Davis' cycle remains a convenient theoretical model for discussing landform development.

Initially, according to Davis, the land surface is gradually lifted above sea level, where it is subjected to erosion activity (see Figure 17–12A). Gradually tiny streams or **rivulets** develop as erosion wears away the land surface at varying rates. In humid regions, the cycle evolves such that rivulets

DAVIS' GEOMORPHIC CYCLE

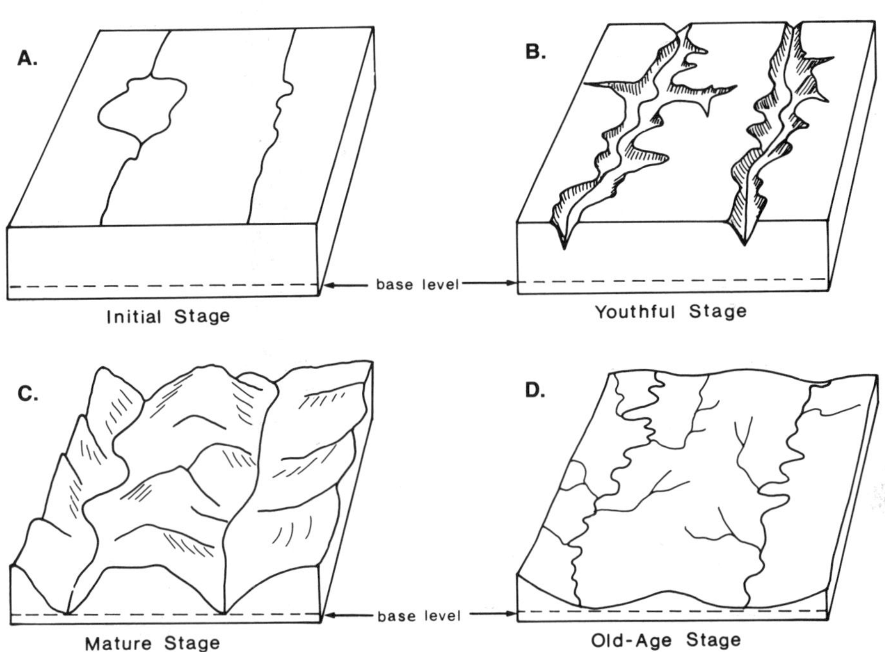

Figure 17–12

develop into distinct stream channels and begin to dissect the landscape. In the youthful stage, the land areas between stream channels, called **interfluves,** are relatively flat and dominate the landscape; but as the youthful stage progresses, interfluves display steeper slopes as they are downcut rapidly (see Figure 17–12B). When downcutting has reached a maximum and lateral cutting begins, the geomorphic cycle advances from the youthful stage to maturity.

At the beginning of maturity (see Figure 17–12C), the difference in elevation or relief between the tops of interfluves and valley bottoms is at maximum. As maturity continues, the stream cuts a wide floodplain that reduces the width of the interfluves. Simultaneously, due to continued weathering and erosion, the height of the interfluves is also reduced. By late maturity, stream valleys are very wide, the number of tributaries is at maximum, and interfluves are eroded into low, gently sloping hills.

The final stage in the geomorphic cycle is called *old age* (see Figure 17–12D). By the time the landscape evolves to this stage, erosion has stripped away nearly all remaining interfluves. Valleys are extremely broad and flat; floodplains are at maximum width; relief is very low; the number of tributaries is reduced into a few major channels; and, except for a few resistant knobs of bedrock called *monadnocks,* the landscape has evolved back to its initial form—that is, a flat plain that lies near or at sea level. Davis termed these theoretical erosional plains **peneplains.** No peneplains exist today. In fact, while some geologists acknowledge the existence of former peneplains, others argue that there never were any peneplains. In any case, it is safe to say that the development of peneplains is very rare because the earth's crust is constantly subjected to tectonic activity and thus rejuvenation is possible during any stage in the geomorphic cycle.

The development of landforms in humid regions just discussed differs from the cycle of development of landforms in dry regions. The geomorphic cycle in dry regions will be explained in Module 19.

Questions

1. What factors can cause changes in stream velocity?

2. How is stream velocity related to stream gradient?

3. Why are semicircular channels best suited for high-velocity streams?

4. What is the difference between stream load and stream capacity?

5. What is the relationship between stream discharge and stream velocity?

6. What are the various ways in which streams erode their channels?

7. What is the relationship between stream velocity and stream competency?

8. How can stream materials be transported? What is the relative importance of each method in transporting debris?

9. What is the relationship between stream deposition and stream velocity?

10. What landform features are formed by stream deposition? Can you explain how each is formed?

11. How can changes in base level influence stream erosion and stream deposition?

12. What are the basic characteristics of a youthful stream? A mature stream?

13. How do meandering streams form a floodplain?

14. How are natural levees formed? What are yazoo tributaries?

15. How can rejuvenation interrupt the stages of growth of a stream?

16. How are stream terraces formed?

17. What are the stages in Davis's geomorphic cycle? Can you describe the nature of the landscape during each stage?

18. Why is there a geologic controversy over the concept of peneplains?

Suggested Readings

LEOPOLD, L. B.; WOLMAN, M.; and MILLER, J. 1964. *Fluvial processes in geomorphology*. San Francisco: W. H. Freeman.

MORISAWA, M. 1968. *Streams: their dynamics and morphology*. New York: McGraw-Hill.

OLIVER, J. E. 1977. *Perspectives on applied physical geography*. North Scituate, MA: Duxbury Press.

Module 18

Landforms from Glacial Action

INTRODUCTION

Within the past 2.5 million years, large masses of ice, called *glaciers,* moved over sections of North America, Europe, and Asia, eroding the landscape and forming new landforms by erosion and deposition. Indeed, the shapes of many of our present landforms are the direct result of previous glacier activity. Today, however, glaciers cover only about 10% of the earth's surface. They are now confined to regions in very high latitudes or to regions with very high mountains where average monthly temperatures are low enough to support a permanent snow cover.

Some meteorologists and glaciologists support the idea that the earth is heading into another ice age. Indeed, at least one meteorologist believes that it is possible that glaciers will advance as far south as Chicago in about 100 years from today. However, while it is true that the average temperature around the earth has declined since 1940, most scientists agree that it is still too soon to accurately predict whether the cooling trend of recent years will continue. In fact, there are almost as many scientists who argue that the earth's atmosphere will begin to warm within 100 years as there are scientists who predict a continuing cooling trend.

Glaciers are of interest not only to meteorologists and glaciologists but also to the government of Saudi Arabia, for they represent a potential source of fresh water. It has been seriously proposed that large sections of the Antarctic ice sheet be separated from the continent and tugged across the ocean by ships to Saudi Arabia, where the ice will melt and provide an estimated several hundred thousand gallons of fresh water. The cost of the project is estimated well into the millions of dollars.

OBJECTIVES

By the end of this module, you should be able to do the following:

1. Explain how a glacier develops from accumulated snowfalls.
2. Describe the flow characteristics of glaciers.
3. Explain the methods by which glaciers erode.
4. Explain the different types of glacially deposited material.
5. List the four major stages of glaciation during the Pleistocene epoch.
6. Explain and describe the various landform features that are attributed to alpine glacier erosion and deposition.
7. Explain and describe the various landform features that are attributed to erosion and deposition by ice sheets.

KEY TERMS

snowfield	col
firn (névé)	hanging valley
crevasse	fiord
plucking	lobe
abrasion	moraine
striations	kettle
scoured rock	drumlin

drift	boulder train
till	esker
erratics	crevasse filling
Pleistocene epoch	kame
cirque	glacio-fluvial
tarn	outwash plains
rock step	lacustrine plain
paternoster lake	underfit stream
arête	calving
horn	iceberg

Now you are ready to begin the audiovisual portion of this module. Select the MEDIAPAK 18 *component(s) and proceed. Following is a topical outline of the audiovisual sequence. You will find this outline helpful for reference and review. After completing* MEDIAPAK 18, *return to this book to perform the exercises.*

OUTLINE

Glacial formation

Glacial flow

Glacial erosion

Glacial materials

Pleistocene epoch

Alpine glaciers
Alpine landforms

Continental glaciers (ice sheets)
Continental landforms

Glacio-fluvial features

Exercises

1. The average depth at which firn develops in a snowfield is ___________ feet.
    A. 50
    B. 200
    C. 15
    D. 100
    E. 150

2. Which zone of a glacier flows most rapidly?
    A. Zone of flow
    B. Zone of fracture
    C. Zone of wastage
    D. Zone of accumulation

3. Scoured rocks and striations are the result of plucking action.
    A. True
    B. False

4. Which of the following correctly lists the glacial stages of the Pleistocene epoch, from oldest to most recent?
    A. Nebraskan, Kansan, Illinoian, Wisconsin
    B. Kansan, Illinoian, Nebraskan, Wisconsin
    C. Wisconsin, Nebraskan, Illinoian, Kansan
    D. Nebraskan, Wisconsin, Kansan, Illinoian
    E. Illinoian, Kansan, Wisconsin, Nebraskan

5. A narrow ridge between two alpine glaciers is called a(n)
    A. esker.
    B. kame.
    C. moraine.
    D. arête.
    E. cirque.

6. The furthest advance of a glacier is usually marked by a high ridge composed of till called a(n)
    A. drumlin.
    B. crevasse filling.
    C. terminal moraine.
    D. outwash plain.
    E. medial moraine.

7. Which of the following features is associated only with ice sheets?
    A. Medial moraine
    B. Arête
    C. Drumlin
    D. Cirque
    E. Ground moraine

8. Which feature is formed by meltwater streams beneath a glacier?
   A. Esker
   B. Arête
   C. Crevasse filling
   D. Drumlin
   E. Fiord

9. Calving is a glacier's action of pulling rocks away from the rock mass over which it moves.
   A. True
   B. False

10. Alpine glaciers sometimes moved over land, cutting valley bottoms below sea level and eventually reaching the ocean. The result of this glacial retreat is a drowned valley called a
    A. paternoster.
    B. fiord.
    C. lacustrine valley.
    D. col.
    E. hanging valley.

Summary

Glacial Formation

Glaciers form from portions of annual snowfalls that do not melt during warm seasons and thus accumulate year after year. Gradually, the accumulated snow builds up to a considerable depth and area sufficient to form a **snowfield.** Snowfields are the source for glaciers. Once snowfields attain a certain depth, the lowest layers of snow are greatly compacted by alternate thawing and freezing. Snow at these layers is compacted into ice granules called **firn** or **névé.** Firn will form when the snowfield reaches an average depth of about 100 feet (30 m). Firn is further modified as the snowfield deepens: Air is squeezed out of the pore spaces between the firn granules, and the firn is compressed into one solid ice mass called *glacial ice.* Glacial ice may be tens of feet thick in small mountain glaciers and thousands of feet thick in the large continental ice sheets.

Glaciers have two major parts (see Figure 18–1). The front edge of the glacier will advance or melt back (retreat) based on whether snow accumulation exceeds melting or vice versa. The front section of the glacier is termed the *zone of wastage,* and the back section is called the *zone of accumulation.* Glacier advance or retreat is largely a function of climate. In the early part of

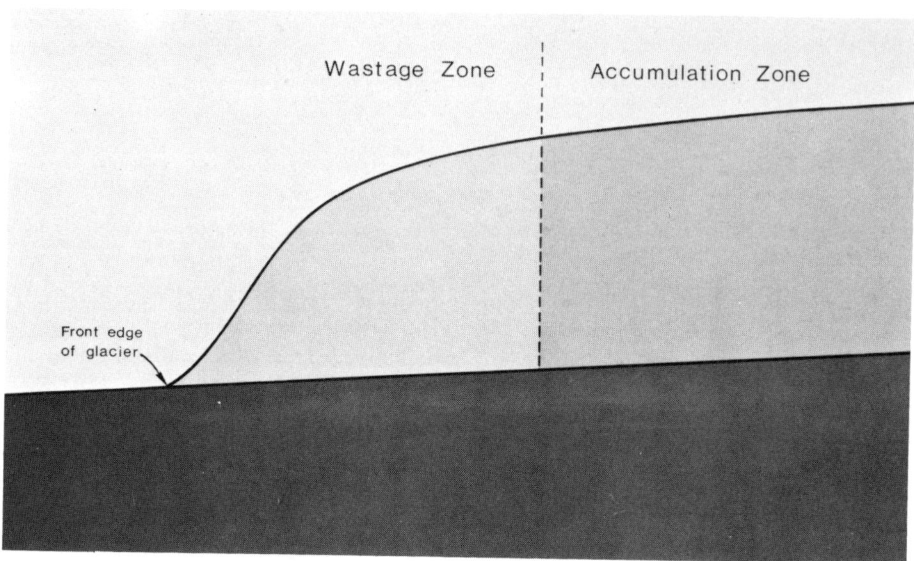

Figure 18–1

the twentieth century, glaciers of the world were generally in retreat. Since 1940, however, most glaciers have stabilized or advanced, which suggests a general cooling of the world's climate.

Glacial Flow

Glaciers generally move very slowly; a few inches or feet per day is a large amount. On occasion, a glacial surge may take place in which the glacier advances several hundred feet within a few days. What causes a glacial surge is still unknown, but it is thought to be related to accumulation of meltwater under the glacier which assists in lubricating the path over which the glacier flows. Long-range studies have shown that glaciers advance more rapidly in their centers than along their edges because friction is reduced in that area.

Glacier movement also varies with increased depth in the glacier. Two zones of movement are identified (see Figure 18–2A). The lower zone, called the *zone of flow,* consists of glacial ice. The lowest layers in this zone are under great pressure and partially melt into a plasticlike consistency. The zone of flow moves at variable rates over objects in its path. The upper zone of the glacier, called the *zone of fracture,* extends to a depth of 100 to 200 feet. It consists of brittle glacial ice, firn, and snow. Unlike the zone of flow, the zone of fracture will not bend as it flows over objects. Instead, this zone cracks apart whenever it bends, forming large breaks at the glacier surface called **crevasses** (see Figure 18–2B). Some crevasses may exceed depths of 50 feet (15 m).

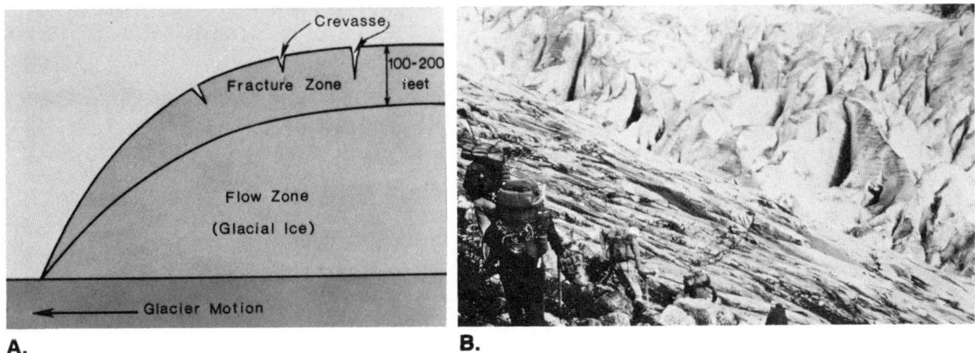

Figure 18–2

Glaciers *erode, transport,* and *deposit* earth materials. Let's discuss glacial erosion and transportation first, and then deposition.

Glacial Erosion

Glaciers erode as the glacial ice flows across surface objects. Accumulations of snow and ice several hundred feet thick exert a tremendous amount of pressure at the bottom of a glacier. As the glacier advances, the plasticlike glacial ice flows over and around any projecting surface material in its path. As the glacial ice strikes a resistant object, the ice is slightly compressed and melts. The meltwater, in turn, flows into the cracks in the projecting material and refreezes, forming a tight bond between the ice mass and the material. As the glacier flows along, the projecting material, which is now frozen to the glacier, is literally ripped free from the main body of surface material and is transported with the moving glacier. This type of glacier erosion is called **plucking.**

A second form of glacier erosion is called **abrasion.** Materials that the glacier has plucked as it flows along act as cutting tools as they are dragged over other materials. Some rock surfaces have been abraded by moving glaciers and thus often display deep grooves and scratches called **striations.** Other rock surfaces have been highly polished by abrasion, if finer materials were involved in the abrasive action. Rock surfaces polished in this fashion are called **scoured rocks.**

Glacial Materials

The rock material eroded by a glacier is eventually deposited as a glacier stagnates or retreats by melting. All glacially deposited material is called **drift.** Drift is subdivided into *unstratified* rock deposits laid down directly by the glacier, and *stratified* rock materials that have been sorted and redeposited by meltwater from the glacier. Unstratified drift, called **till,** is a jumbled mixture of all sizes of particles deposited by a glacier. Stratified drift is glacial deposits that have separated into distinguishable layers based on size; boulders form one layer, smaller rocks a second layer, gravel a third, and so on. Boulders in glacial deposits are called **erratics** because generally they are unlike the underlying bedrock at the location where they are lying.

Glacial Drift

The term *glacial drift* was first introduced by C. Lydall in 1840. The term dates from the time when it was generally thought that the large boulders and other rock deposits found scattered across much of Europe were the result of deposition by the biblical Great Flood. Later glacier studies provided evidence that moving ice was capable of transporting and depositing large boulders. By the late 1870s, support for Noah's flood as an explanation for the scattered rock material had been replaced by support for the theory of moving glaciers. Nevertheless, the term *drift* remains part of glacial terminology.

Pleistocene Epoch

Glaciers have covered portions of the earth's surface to varying extent throughout history. They have occupied portions of Europe, the midwestern United States, Canada, and the northwestern section of the Soviet Union, and they have extended from Antarctica and Greenland into the surrounding oceans. At their extreme, glaciers covered about 27% of the earth's surface. Glacial activity reached its peak during the **Pleistocene epoch,** or Ice Age, which began about 2.5 million years ago. The Ice Age ended about 10,000 years ago as glaciers began to recede to their present locations. Today, glaciers are located in various high mountain areas and on the continent of Antarctica and in Greenland.

The Pleistocene epoch saw the advance and retreat of glaciers at least four times. Each major period of advance is termed a *glacial stage*. In North America, the four major stages were as follows:

Nebraskan stage	1,000,000 years ago
Kansan stage	750,000 years ago
Illinoian stage	300,000 years ago
Wisconsin stage	10,000 years ago

Between the major stages of advance were interglacial periods of general climatic warming and glacial retreat.

Two categories of glaciers are recognized: *alpine glaciers,* which occupy individual mountain valleys; and *continental glaciers,* or ice sheets, which cover entire continents or portions of continents. The primary difference between the two is scale or size. Alpine glaciers may be several hundred feet thick, while ice sheets may be more than 10,000 feet thick. The Antarctic ice sheet, for example, measures slightly in excess of 13,000 feet at its thickest point. Alpine glaciers are generally confined to mountain valleys and tend to sharpen the preglacial landforms; continental glaciers, on the other hand, overwhelm the landforms of a large portion of a continent and generally streamline those landforms. We will discuss alpine glaciers first and then continental glaciers.

Alpine Glaciation

Today, alpine glaciers, also called *valley* glaciers or *mountain* glaciers, are located today at all latitudes where mountains are high enough to provide cold temperatures for year-round snow. As alpine glaciers advance downslope from the uppermost sections of valleys, they erode both the valley sides and the valley bottoms. As a result, they make several modifications in the original valley. The backwall of rock against which the head of the glacier rests is constantly weathered and eroded by plucking and frost action. By weathering and erosion, the backwall is gradually hollowed out into a semicircular, amphitheaterlike form called a **cirque** (see Figure 18–3). Once a glacier has

Figure 18–3

melted, cirques may contain small lakes called **tarns.** In some alpine valleys, a series of cirques may form down the valley. The result is a series of carved **rock steps** that may contain small lakes after the glacier retreats. A series of these small lakes down the valley are called **paternoster lakes** because they resemble beads on a rosary. Paternoster lakes are features of valleys composed of rock of unequal resistance. In valleys with uniform bedrock, the glacier carves the valley floor into a smooth trough.

Each separate valley in a mountain range may contain an alpine glacier. As two adjacent alpine glaciers erode the ridge of bedrock that separates them, the ridge may be eroded into a very thin, sharply edged feature called an **arête** (see Figure 18–4). In a similar fashion, if two or more cirques erode the same headwall from different directions, the remaining rock feature may be a sharp peak called a **horn.** The Matterhorn in Switzerland is a well-known example. A col is another feature that is the result of erosion of the same headwall by two or more glaciers. When erosion completely strips away the headwall, a pass is created through the mountain range; this pass is called a **col.**

Large alpine glaciers may be joined by smaller tributary glaciers, just as stream tributaries join a river. Because of their mass, large glaciers are capable of gouging the valley bottom to a greater depth than smaller tributary glaciers. After the glaciers melt, therefore, there remain one deep main valley and also several tributary valleys whose floors lie high above the main valley floor. The smaller tributary valleys are termed **hanging valleys,** many of which are characterized by waterfalls (see Figure 18–4).

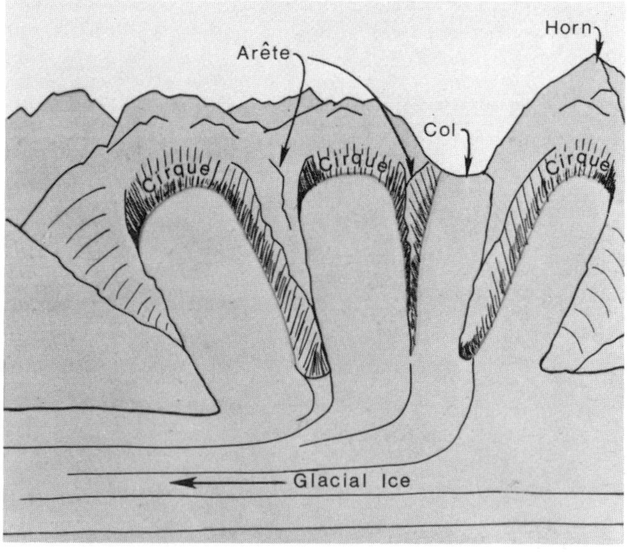

Figure 18–4

Where alpine glaciers occupy mountain regions near seacoasts, glaciers can advance some distance out to sea, eroding the valley bottoms below sea level. When the glacier melts, seawater partially fills the eroded valleys. Glacial valleys filled with seawater are called **fiords.** Fiords extend inland along the coasts of Alaska, British Columbia, Norway, Greenland, Labrador, New Zealand, and Chile.

An indication of glacial activity in mountain regions is the profile of the mountain valleys (see Figure 18–5). Most *mountain* stream valleys have a V-shaped or modified V-shaped profile, with a main stream that runs about at the bottom of the V. Glaciers move through a valley and erode both the bottom of the valley and the sides of the valley up to the height of the ice thickness. Consequently, *glaciated* valleys are carved into U-shaped channels.

Figure 18–5

As a glacier advances, it erodes rock materials from the valley bottom that it occupies. Rocks also collect on the top of the glacier as they are broken loose from the valley walls by frost action and tumble down and collect on the glacier surface. New rock materials are constantly carried forward to the front edge of a glacier, even as a glacier retreats, for the glacial ice within a glacier advances even as the glacier retreats. The only reason that a glacier retreats is that new ice is not being carried forward as fast as the old ice melts away. Eroded rock materials continue to be transported forward in a process that acts like a conveyor belt. If the leading edge of the glacier remains stationary for a period of time, the amount of rock material carried by the ice to the leading edge of the glacier can eventually build into a very thick deposit. A high ridge, called a **moraine,** is built in this fashion. Moraines are the most widespread feature of glacial deposition and are characteristic of both alpine and continental glaciers. Moraines may be several hundred feet high and many miles long.

Moraines that mark the point of furthest glacier advance are termed *terminal moraines* or *end moraines* (see Figure 18–6). As glaciers retreat from the terminal moraine, rock material continues to be transported forward in the ice. Moraines will not build to any great size unless the glacier stagnates for a sufficient period of time. If the ice does not stagnate, materials are deposited in an uneven mass, just as they had been collected and transported by the glacier. This uneven mass of glacial debris is called *ground moraine.* Deposits of ground moraine usually form a very rolling terrain sometimes referred to as *swell and swale topography.*

If a glacier stagnates for a period of time during a period of general retreat, a moraine may build up at the temporary position of the glacier's leading edge. These moraines, called *recessional moraines,* lie in back of and parallel to the terminal moraine. Like ground moraines, recessional moraines are features of both alpine and continental glaciers, but they are usually less developed than terminal moraines. It is not uncommon for several recessional moraines to form as a glacier stagnates several times periodically during its general retreat.

Two types of moraine are features only of alpine glaciers. Rock debris collects along the edges of alpine glaciers as valley walls are eroded. Ridges of rock debris build up and are deposited when the glacier melts; these ridges are called *lateral moraines.* As a tributary glacier flows into a main glacier, the glacial debris carried along the edges of both glaciers joins together and forms one large ridge called a *medial moraine* (see Figure 18–6). Medial moraines form down the central part of the main glacier and usually form a low, straight ridge down the main valley after the glacier melts. Medial moraines are often buried by ground moraine, and, for that reason, they are difficult to identify.

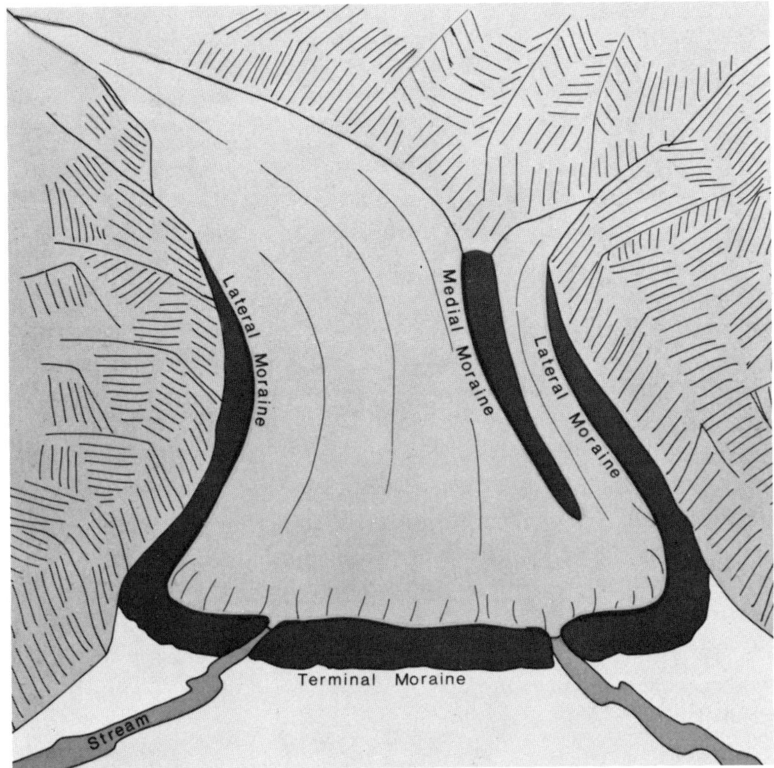

Figure 18–6

Continental Glaciers (Ice Sheets)

Many of the features of alpine glaciers are also features of ice sheets, except the features develop to a larger scale in ice sheets. Terminal moraines, recessional moraines, and ground moraine are features of both types of glacier.

Continental glaciers advance from a central source region in a series of fingerlike projections called **lobes** (see Figure 18–7). Each separate lobe may cover tens of thousands of square miles, and each has its own special ability to erode, transport, and deposit materials. Variations from one glacier lobe to the next can result in a significant variation in the type of landform modification produced by those lobes. Some regions are stripped clean of all surface materials, with exposed, scoured bedrock left behind, while other areas are buried beneath hundreds of feet of drift and display a complex pattern of hills and valleys.

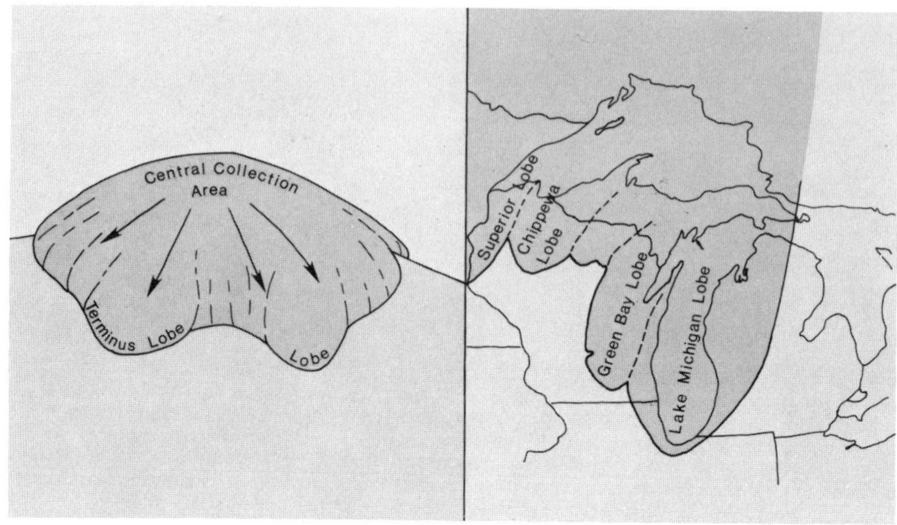

Figure 18–7

As mentioned earlier, moraines of many types are features of both ice sheets and alpine glaciers. One moraine type common only to ice sheets, however, is the interlobate moraine, a ridge formed where moraines along the sides of two ice sheet lobes join together.

As ice sheets gradually retreat, tremendous quantities of drift are deposited. At the time of melting, large blocks of ice are sometimes buried beneath till deposits. Once these ice blocks melt, the till collapses and a depression is formed in the till deposits. These depressions are called **kettles** (see Figure 18–8). Kettles that contain water are termed *kettle lakes*. Kettles are relatively common features in the moraine deposits of ice sheets.

Drumlins are elongated hills formed by deposits of till under ice sheets (see Figure 18–9). Although the origin of drumlins is still being debated, their shapes are easily distinguishable on the landscape. Drumlins are blunt and steep-sloped at one end and have a gradual slope at the opposite end. From an airplane they resemble an overturned teaspoon without a handle. The blunt end of the drumlin points to the direction from which the glacier advanced. Drumlins range from 50 to 200 feet in height and are about ¼ mile to ½ mile long and about ¼ mile wide. They appear less frequently than many other glacial features. Where they are located, however, they tend to appear in bunches that are termed *drumlin fields*. Major drumlin fields appear in portions of the states of New York, Wisconsin, Michigan, and Minnesota, and in Nova Scotia.

Glaciers several thousand feet thick have tremendous erosive power. They can pick up large boulders, many as large as a house, and transport them several hundred miles before depositing them. On occasion, many boulders may be deposited in an arrangement that appears as a long line. These deposits

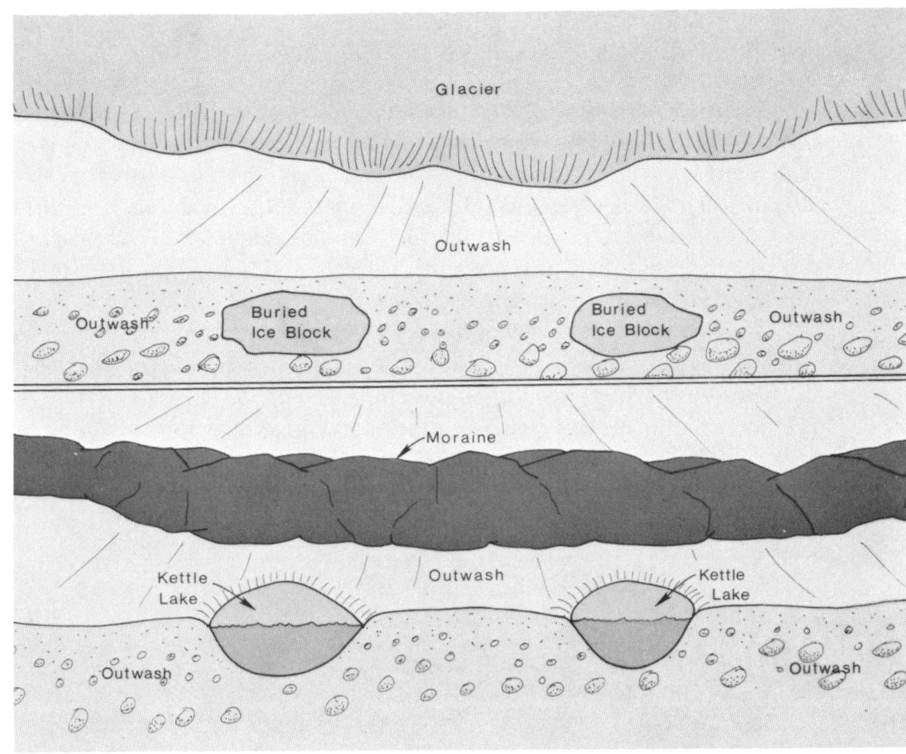

Figure 18–8

Figure 18–9

are called **boulder trains.** Boulder trains are good indicators of the direction of glacier flow.

Ice sheets release large volumes of meltwater that channel into streams that flow over, through, and below the glacier. These meltwater streams carry with them much of the glacial drift that the glacier was transporting. When the stream volumes decrease as glaciers retreat, large quantities of glacially eroded materials are deposited and form various features. All meltwater features contain stratified drift because water is a selective erosion agent that separates material into distinctive sizes.

One meltwater feature that forms in streams that flow beneath the glacier when it stagnates is called an **esker.** After a glacier retreats, eskers appear on the landscape as long, winding ridges composed of stratified drift. They range from only a few feet to as many as nine hundred feet in height, and they snake back and forth across the landscape much like stream channels (see Figure 18–10). Many eskers today have been partially destroyed by sand and gravel operations that tap the stratified deposits.

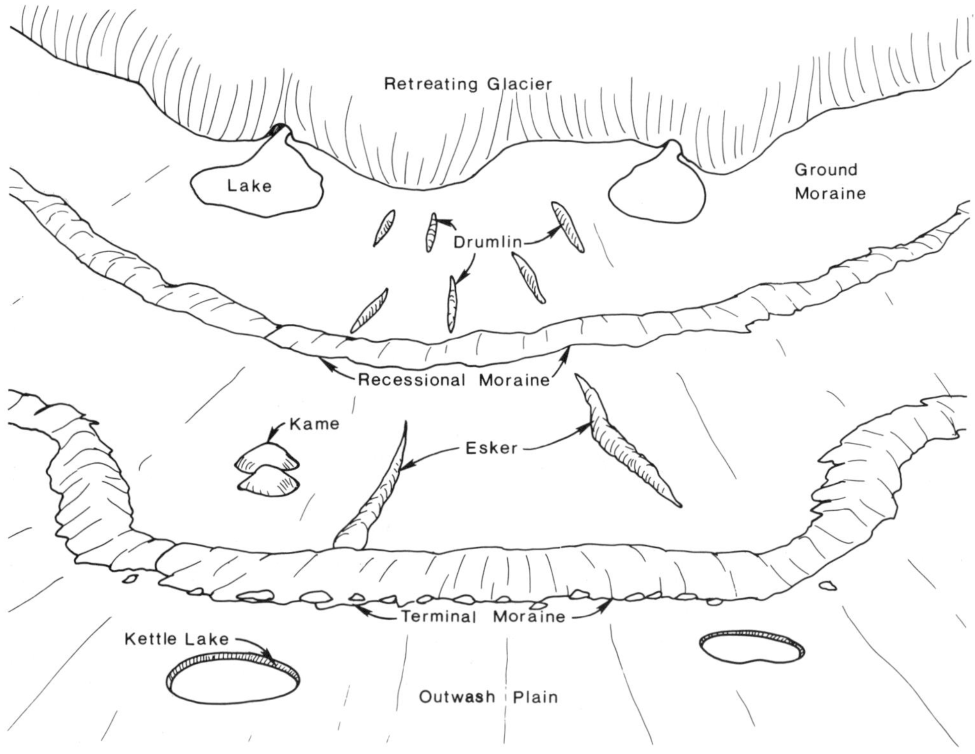

Figure 18–10

Two other features form when ice sheets stagnate: crevasse fillings and kames. Meltwater flowing on the surface of the glacier flows into crevasses and washes drift into the crevasse. When the glacier melts, the material that flowed into the crevasse stands as an elongated, straight ridge, which generally lies at right angles to the direction of glacier flow; this feature is called a **crevasse filling.** In a similar fashion, low, steep-sided hills, called **kames,** form as meltwater washes rock and earth debris through holes in the glacier. After the glacier retreats, the kames often appear as nearly perfectly round hills of varying size (see Figure 18–10). Crevasse fillings and kames may both exceed 100 feet in height. Crevasse fillings are generally 50 to 150 feet long and about 50 to 60 feet wide. Kames may have diameters of several hundred feet at their base.

Glacio-Fluvial Features

Meltwater flowing off the leading edge of a glacier is capable of carrying glacial material beyond the terminal moraine. As a result, glacier meltwater features form beyond the terminal moraine; these features are called **glacio-fluvial features.** Several landforms are formed from glacio-fluvial action. For example, **outwash plains** are features of both ice sheets and alpine glaciers, although they are best developed in association with ice sheets. They are gently sloping surfaces composed of coarse gravels and sands that were transported and deposited by meltwater as it ran off the front edge of the glacier beyond the terminal moraine (see Figure 18–10). Outwash plains are usually stratified, with the largest rocks deposited closest to and the finest particles farthest from the glacier. Close to the terminal moraine, ice chunks may be carried along with meltwater, only to be buried by outwash. When the ice chunk melts, the outwash sinks and forms a kettle. Outwash plains with kettle holes are called *pitted outwash plains.*

The large volume of meltwater from glaciers formed some very large glacial lakes, most of which have dried up since the ice sheets retreated. Glacial Lake Agassiz, for example, was centered in Manitoba about 10,000 years ago and covered nearly 200,000 square miles—twice the area of the present Great Lakes. The Great Lakes themselves were gouged by glacial erosion, and during the Wisconsin stage of glaciation, they occupied considerably more area than at present. The imprint of former glacial lakes remains on the landscape in the form of **lacustrine plains,** flat plains that are the lake beds of the former glacial lakes. Many of these former lake beds are now utilized for various types of agriculture (see Figure 18–11).

Meltwater stream channels flowing from ice sheets may be very broad. They continue to flow with large volumes until the glacier retreats. Once the glacier retreats, the volume of water decreases substantially, and the stream is

Figure 18–11

reduced in size. The smaller stream is termed an **underfit stream** because the original channel is much wider than the present stream could possibly erode even at flood stage. Meltwater stream channels often become choked with sediment when the volume of water decreases. As the smaller stream cuts a channel through this sediment, terraces can form along the banks of the smaller stream.

Today, ice sheets can be seen only in Antarctica and Greenland. Where the ice sheets meet the sea, the contrast in temperature between ice and seawater and the vibrations from wave action cause the ice sheets to break apart along their edges. This breakup action is called **calving.** As a result of calving, huge chunks of ice, called **icebergs,** float out to sea and can pose a hazard to ocean-going ships.

Questions

1. What percent of the earth's surface is covered by glaciers today?

2. What has been the general motion of glaciers since 1940?

3. How does the flow of a glacier vary with depth?

4. What are the two methods of glacial erosion?

5. What is the relationship between drift, stratified drift, and till?

6. What are the four major stages of glaciation during the Pleistocene epoch, and how many years ago did each stage end?

7. Why are the profiles of glaciated mountain valleys usually U-shaped, while those of nonglaciated valleys are generally V-shaped?

8. How do cirques form?

9. What is meant by the statement, "Glaciers advance even while retreating"?

10. How many different types of moraine are there? How is each type related to the glacier?

11. What are the various glacial meltwater features that form behind the terminal moraine?

12. List the various glacio-fluvial features. How is each feature formed?

13. What causes calving?

Suggested Readings

EMBLETON, C., and KING, C. A. M. 1968. *Glacial and periglacial geomorphology.* Edinburgh, Great Britain: Edward Arnold Publishers Ltd.

FLINT, R. F. 1971. *Glacial and quaternary geology.* New York: John Wiley & Sons.

OLIVER, J. E. 1977. *Perspectives on applied physical geography.* North Scituate, MA: Duxbury Press.

SCHULTZ, G. 1963. *Glaciers and the Ice Age.* New York: Holt, Rinehart and Winston.

Landforms from Wind and Waves

INTRODUCTION

Although *wind* is the least effective of the major erosion agents, it is capable of modifying existing landforms, particularly in arid regions where vegetation is sparse. Landforms formed by wind activity are called *eolian* landforms.

Waves, on the other hand, are very effective erosion agents, and the coastlines of the world are in constant battle against the sea. Waves and currents strip land away from some shore areas and deposit the eroded material along other stretches of coastline. People have used coastal waters for fishing and have constructed major cities along coasts to take advantage of natural harbors. In addition, they have greatly modified coastlines by building breakwaters and by dredging harbors. In some places, human attempts to combat erosion by the sea have failed completely, whereas in other regions, large areas of land have been reclaimed from the sea through the construction of dikes and landfill. Human efforts to change coastlines have not always been successful, however. Attempts to construct some harbors, for example, have accelerated deposition where it was not supposed to happen, which in turn triggered rapid erosion where previously erosion was not a problem.

OBJECTIVES

By the end of this module, you should be able to do the following:

1. Explain and describe the two methods of wind erosion and the landforms associated with each method.
2. Describe the various types of sand dunes.
3. Explain and describe the evolution of landforms in arid regions, and describe the landforms associated with this cycle of development.
4. Explain and describe the features associated with an eroding plateau in arid regions.
5. Explain the action of a breaking wave and the relationship between waves and coastal landforms.
6. Explain the action of longshore currents.
7. Explain and describe the landforms associated with erosion by waves and currents.
8. Explain and describe the landforms associated with deposition along irregular and relatively straight coastlines.
9. Describe the features associated with submerged, emerged, and neutral coastlines.

KEY TERMS

deflation	butte
abrasion	hamada
deflation hollow (blowout)	continental shelf
desert pavement (reg)	sea cliff
pedestal rocks	wave-cut notch
ventifact	undertow
saltation	wave-cut bench

slip face	wave-built bench
erg (sand sea)	marine terrace
loess	longshore current
ephemeral stream	sea stack
wadi (wash, arroyo)	headlands
pediment	bar (spit)
playa	tombolo
pediplain	barrier beach
inselberg	ria shoreline
mesa	estuary

Now you are ready tobegin the audiovisual portion of this module. Select the MEDIAPAK 19 *component(s) and proceed. Following is a topical outline of the audiovisual sequence. You will find this outline helpful for reference and review. After completing* MEDIAPAK 19, *return to this book to perform the exercises.*

OUTLINE

Wind erosion
Deflation
Abrasion

Wind deposition
Sand dunes
Loess

Arid landform evolution
Plateau modification in arid regions

Coastal landforms
Waves and currents
Erosional features
Depositional features
Classification of shorelines

Exercises

1. Which sand dune is an elongated ridge that runs parallel to the wind direction?
   A. Parabolic
   B. Longitudinal
   C. Barchan
   D. Transverse

2. Which arid landform feature is a byproduct of deflation?
   A. Pediment
   B. Playa
   C. Reg
   D. Pedestal rock
   E. Ventifact

3. Which of the following features is most often mistakenly identified as an alluvial fan?
   A. Playa
   B. Erg
   C. Reg
   D. Hamada
   E. Pediment

4. If a plateau is composed of easily eroded material, erosion will produce a landform called
   A. a playa.
   B. badlands.
   C. a mesa.
   D. a gorge.
   E. a canyon.

5. Breakers at high tide erode into the base of rock cliffs and produce a feature called a(n)
   A. wave-built terrace.
   B. sea stack.
   C. wave-cut notch.
   D. bar.
   E. estuary.

6. Abraded materials collect on the downcurrent side of headlands.
   A. True
   B. False

7. Barrier beaches are most characteristic of _________________________ coastlines.
   A. emerged
   B. submerged
   C. neutral
   D. compound
   E. ria

8. A shoreline that has exposed marine terraces is a(n)
   A. ria shoreline.
   B. submerged coastline.
   C. emerged coastline.
   D. compound coastline.
   E. neutral coastline.

9. Which coastline of the United States is a submerged coastline?
   A. Atlantic coast
   B. Pacific coast

10. A wave-cut bench is cut out of bedrock as a sea cliff erodes.
    A. True
    B. False

Summary

Wind Erosion

Wind erosion is accomplished by deflation and abrasion. **Deflation** is the process whereby the wind picks up tiny particles off the ground and carries them to another location; deflation is, therefore, primarily a *transporting* action. It is most active where the ground material is loosely consolidated and the ground is exposed to the wind due to lack of a vegetation cover.

Deflation most often involves clay and silt particles that can be moved with even moderate wind velocities. The larger the particles involved, the greater must be the wind velocity to move them. Deflation forms few landforms directly. A **blowout,** or **deflation hollow,** is a shallow depression formed in part by deflation. As depressions fill with water during intermittent rain showers in dryland areas, sediment collects in the water. When the water in the depression evaporates, the sediment settles on the floor of the depression and cracks as it dries out; the sediment cracks into thin scales that the wind carries away, creating a blowout. Overgrazing by cattle helps to form blowouts as vegetation is trampled or eaten and bare ground is exposed. Blowouts are most common in the semiarid regions of the Great Plains. They range from 1 foot to 30 feet in depth, and from about 10 feet to over a mile in diameter.

Where deflation is most active, all the fine particles may be removed from an area. As a result, a surface of loose rock pebbles and gravel remains. As the wind jostles these particles back and forth, they eventually fit tightly together and form a relatively flat surface called **desert pavement** or **reg.**

Salts, drawn to the surface by evaporation, often help to cement the pebbles together.

Abrasion, the second form of wind erosion, is the process whereby the particles transported by the wind bombard solid objects. Abrasion is particularly effective on objects that extend the first several feet above the general land surface. The constant smashing of windblown silt and fine sands against solid rock is capable of gouging holes in the rock. Arches may be formed by this action. Irregularly shaped rocks, called **pedestal rocks,** can have very narrow bases and broader tops due to abrasion (see Figure 19–1A.) In areas where the wind blows consistently from one direction, rocks may be flattened on one side by abrasion. These angular rocks are called **ventifacts** (see Figure 19–1B).

Strong winds in desert and semiarid regions are capable of picking up huge volumes of sand and dust creating a sandstorm or a dust storm. *Dust storms* may contain several million tons of dust; they may transport dust particles several thousand feet up and as far as 2500 miles. *Sandstorms* lift sand particles to heights of about 6 feet. Where sandstorms are common, wood-frame homes and telephone poles often require metal sheathing around the lower several feet to protect the wood from being cut away by blowing sand.

Wind Deposition

Wind-eroded materials are deposited when wind velocity declines. Sand deposits accumulate into hills called *sand dunes*. Sand dunes vary greatly in size; they can be a few feet to hundreds of feet in height and can extend more than a mile across at the base. Dunes can change shape and size, and they can move if they are not anchored by vegetation.

Dunes are not exclusively desert features; they are also common along windward, low, sandy coastlines. Dunes associated with beaches are confined to areas within 1 or 2 miles from the shoreline. Both dunes along coasts and those in deserts are derived from particles originally deposited by water, and both have limited vegetation covers. Consequently, some of the best-developed dunes are located along desert coastal regions, as in the Atacama Desert in Chile and the Namib Desert in southwest Africa.

Several dune types form based on the characteristics of the wind and the supply of material. One dune type is called a *barchan dune* (see Figure 19–2). Barchan dunes characteristically have a half-moon shape, with the pointed ends of the crescent pointing downwind. One side of the barchan has a steep slope and the opposite side a gradual slope. The gradual slope faces into the wind, and the steep slope, called the **slip face,** lies on the inside of the crescent. Barchan dunes are usually found in regions where sand is sparse and

Figure 19–1

winds blow from a fairly constant direction. Several barchans can travel
downwind as a chain separated by desert pavement.

Parabolic dunes are a second major dune category (see Figure 19–2).
Parabolic dunes look much like barchans except that the points of the crescent
aim opposite to the direction of the wind. Parabolics also differ from barchans

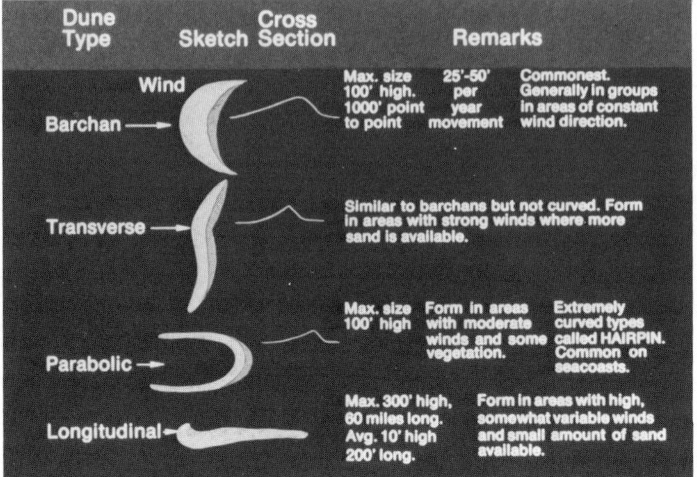

Figure 19–2

in that the steepest slope of the dune is on the windward side. At the foot of the steep slope, inside the crescent, a blowout often develops. Parabolic dunes are more common than barchans where the sand supply is moderate and winds are slightly more variable in direction. Vegetation often anchors parabolics, particularly along coastlines.

A third type of dune, called a *longitudinal* or *seif dune,* is a long ridge of sand that lies parallel to the wind direction (see Figure 19–2). Longitudinal dunes are most common in regions of limited sand and strong, variable winds. They can attain heights of several feet and may be several miles long.

A fourth dune type, called a *transverse dune,* is in contrast to longitudinal dunes (see Figure 19–2). Transverse dunes lie at right angles to the wind direction and are features of regions with an overabundance of sand; these regions are called **sand seas** or **ergs.** It is a common misconception that most desert landscapes are ergs. In fact, only about one-quarter of all deserts contain abundant sand.

Dunes are not simple to classify because of the great variations in appearance that can develop due to changing wind directions, variable volumes of sand, and changes in vegetation cover.

Wind-deposited silt, called **loess,** can accumulate to great depths. In sections of mainland China, for example, loess deposits over 100 feet deep are common, and in some areas, they are 300 feet deep. Loess tightly compacts when it settles, but it is extremely soft material and is easily eroded if not stabilized by a vegetation cover. In North America, loess deposits are located in the upper Midwest. The loess in North America is thought to be derived from glacial deposits located in the northern Great Plains and northern Great

Lakes states. Loess is well suited to the cultivation of grains. Highly productive agricultural regions on loess soils are located in Iowa, Illinois, Kansas, and Nebraska. Other major loess deposits in the world include northern mainland China, the Argentine Pampas, and the southern USSR.

Landforms of Arid Regions

The landforms of arid regions evolve differently in appearance from the landforms of humid regions, despite the importance of running water as an erosion agent in both humid and arid regions. Arid regions generally lack a full vegetation cover. Thus, despite the lack of rainfall, sporadic heavy downpours of rain can rapidly erode exposed bare ground. In addition, wind erosion is most effective where vegetation is sparse.

Streams in desert regions may flow only a short distance before they dry up, at which point stream material is deposited in the stream channel. Such streams that flow only when it rains are called **ephemeral streams.** Stream channels that contain ephemeral streams are characterized by steep, vertical sides and flat bottoms. During a rainstorm, an ephemeral stream erodes vertically and rapidly to form steep, vertical banks. When the stream stops flowing, sediment is deposited in place in the channel, partially filling the channel and forming a flat bed. A steep-sided, flat-bottomed stream channel in an arid region is called a **wadi,** a **wash,** or an **arroyo.**

In arid mountainous regions, ephemeral streams emerge near the base of a mountain and generally spread out, eroding out of bedrock a slope that gently inclines to the valley bottom. Such an erosional slope is called a **pediment.** Pediments are often covered with a thin veneer of alluvium derived from stream deposition. Pediments, therefore, are often mistaken for alluvial fans, which are slopes built entirely of alluvial deposits where mountain streams in humid regions reach valley bottoms (see Module 17).

Depressions between highlands in humid areas fill with water and form lakes, but in arid regions, any water that collects in depressions evaporates. Thus, lakes in depressions in arid regions are usually only temporary and are called **playas.** Playas are generally shallow, and their water levels fluctuate greatly; these lakes often disappear completely for long periods. Playas do not drain into rivers or streams; consequently, salts drawn to the surface by evaporation tend to collect in playas and cause the water to be very salty, often saltier than the oceans.

The lack of steady stream erosion in arid regions means that eroded material is not transported very far from its source. As upland regions are gradually eroded away, the eroded material collects and slowly fills the adjacent lowland areas. After long periods of time, the basins fill completely, and uplands erode away except for a few resistant remnants called **inselbergs.** The landscape then appears as a rolling plain composed of deposited materials, with inselbergs that protrude above the plain. This depositional plain has been termed a **pediplain.** Pediplains are analogous to *peneplains,* a term applied by Davis to old-age depositional plains in humid regions (see Module 17). Pediplains are relatively common features, unlike peneplains which are at best exceedingly rare.

In some arid regions, the landforms are features of an eroded plateau. A plateau is an elevated flat plain. The degree of erosion of a plateau in arid regions is largely dependent on the bedrock. If the bedrock is easy to erode, such as shale, the plateau will be dissected into a maze of steep-sided gullies separated by narrow ridges. This type of landform is referred to as *badlands* (see Figure 19–3). On the other hand, where plateaus are capped by resistant rock layers, the number of dissecting stream channels will be limited. These few channels, however, often cut vertically deep into the plateau and form narrow gorges or wider features called *canyons*.

Figure 19–3

Wind erosion on plateau regions may remove most of the sand and other finer surface materials, thereby exposing the bedrock surface. Flat plateaus of exposed bedrock in arid regions, called **hamadas,** are common features in portions of the Sahara Desert.

Plateaus are gradually eroded by stream activity, wind activity, and mass movement (see Figure 19–4). Sections of a plateau can be separated from the main plateau by erosion, forming individual flat-topped hills, called **mesas.** Mesas, in turn, can erode into smaller, isolated, resistant knobs of rock, called **buttes.** Plateaus, mesas, and buttes are common features in many sections of

Figure 19–4

the southwestern United States, such as Canyonlands National Park in Utah, the Grand Canyon in Arizona, and Mesa Verde National Park in Colorado.

Coastal Landforms

A shoreline is marked by the location of sea level in relation to the adjoining land surface. The edge of the land, however, extends some distance out to sea below sea level. Generally, the submerged outer margins of the continents are gently sloping plains called **continental shelves.**

Continental shelves vary in horizontal extent. In some places, the shelves extend up to 100 miles out to sea, at which distance the sea depth reaches about 600 feet. At the edge of the shelves, the ocean depth quickly increases to several thousand feet. The North Sea and the Bering Strait are nearly totally underlain by continental shelf. In contrast, the coasts of Africa and the west coasts of North America and South America have only very narrow continental shelves. Continental shelves are important as major locations for commercial fishing. The Grand Banks off the east coast of North America, for example, has long been one of the most productive fishing regions in the world.

Coastal landforms are produced by waves and currents along the world's shorelines. Waves move in a circular orbit whose vertical axis is equal to the height of the wave (see Figure 19–5A). When a wave nears a shore, contact

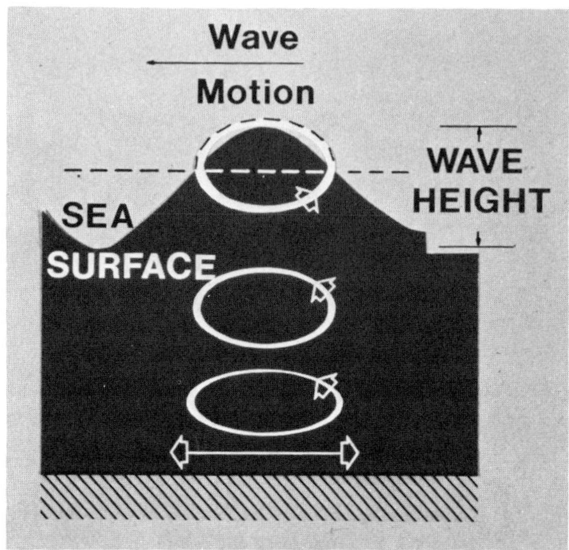

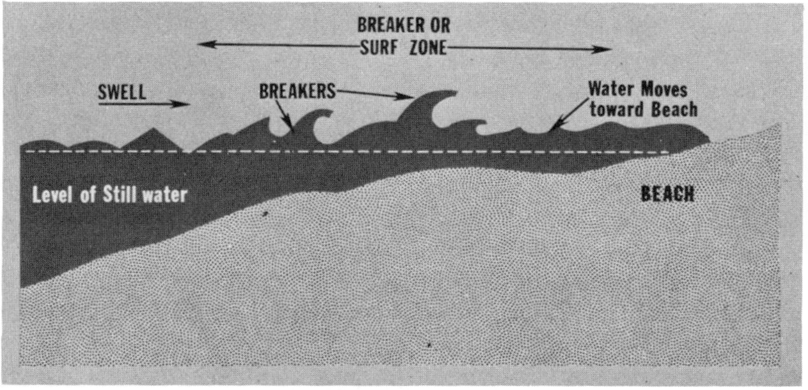

Figure 19–5

between the bottom of the wave and the bottom of the shore causes friction, which reduces the velocity of the bottom of the wave. Consequently, the top of the wave leans toward the shore ahead of the center axis of the wave. The wave becomes higher and narrower until its crest, which leans most toward the shore, is no longer supported by the body of the wave from below. As a result, the wave collapses, producing a breaker (see Figure 19–5B). Once a breaker strikes the shore, the water in the wave flows up the shore or beach some distance above sea level.

As long as waves maintain a nearly round form, they possess little erosive power because they have a low forward velocity. A breaker, on the other hand, has tremendous erosive power as it crashes onto the shore. A breaking wave moves tons of water against the land with pressures in excess of 200 lb/ft². Breakers have their greatest force at high tide when friction with the shore bottom is at a minimum. **Sea cliffs** are eroded most actively at high tide because of the active breakers. Waves breaking at the base of sea cliffs can cut a **wave-cut notch** and maintain the steep face of many cliffs.

The water in a breaking wave flows up the beach and flows back to the sea under the influence of gravity. As it flows back to sea, it moves under succeeding waves and helps upset them. This seaward motion of water beneath landward-moving waves is called the **undertow.** The larger the waves, the greater is the undertow. Large amounts of eroded materials can be transported to the sea by undertowing. Undertows can be powerful enough to pull even the best swimmers out to sea.

In addition to erosion by waves, coastlines are also modified by currents. Undertows are actually one form of current. Another type of current, called a **longshore current,** develops where waves break at an oblique angle to the shore (see Figure 19–6). As waves break against the shore at an oblique angle, the water flows up the beach at the same angle. When the water loses its momentum, gravity pulls the water straight down the beach to the sea, not at an oblique angle. Thus, longshore currents move in a zigzag fashion down

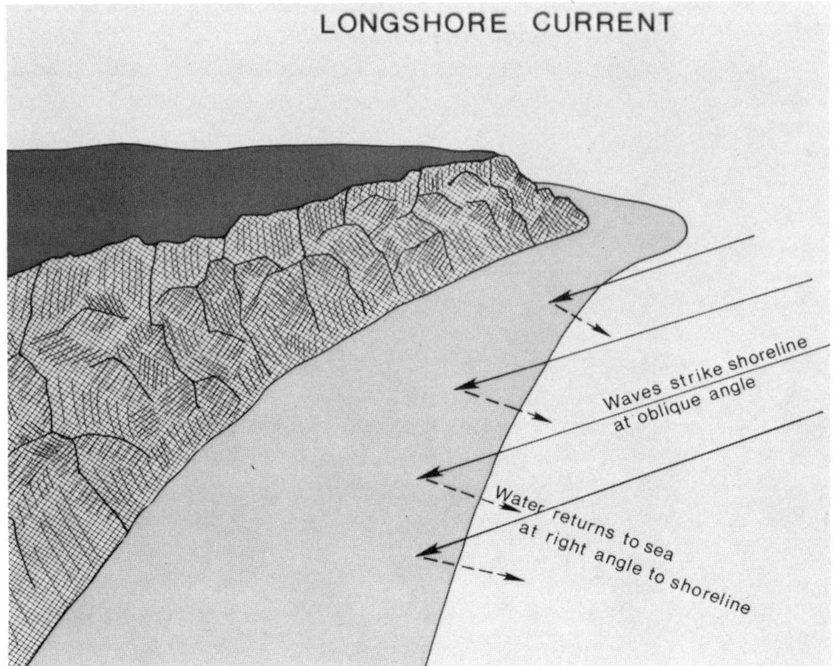

Figure 19–6

the shore. Loose particles and other objects on the beach are transported in a similar fashion. Some objects may be transported hundreds of miles along a coastline by longshore currents.

Where waves erode the shore, a **sea cliff** can form. Sea cliffs continue to retreat inland as waves erode away the base of the cliff (forming a **wave-cut notch**) until the base is eroded sufficiently so that the overlying rocks collapse due to the absence of support from below. As a sea cliff recedes, a **wave-cut bench** forms (see Figure 19–7). Wave-cut benches are smoothed and lowered by continual abrasion as waves and currents drag eroded sea-cliff material across them. If eroded material is abundant, portions of that material may fall over the seaward margin of the wave-cut bench and form a **wave-built bench.** Together, the wave-cut bench and the wave-built bench form what is termed a **marine terrace.** Generally, wave-cut benches and marine terraces are relatively smooth features, but resistant rock remnants occasionally may protrude above the general surface of a bench or a terrace. These isolated resistant rocks are called **sea stacks.**

Coastlines change in appearance not only due to erosion but also due to the *deposition* of materials derived from the erosion of the coast. Abraded materials along a shore are in almost constant motion, although their locations may change only slightly. The rate at which abraded materials are transported and the distance that they travel are dependent, for the most part, on their size, the force and direction of currents, and the shape of the coastline. Sections of the coastline that project into the sea are called **headlands.** Abraded materials accumulate on the upcurrent sides of headlands and are stripped away from the downcurrent sides.

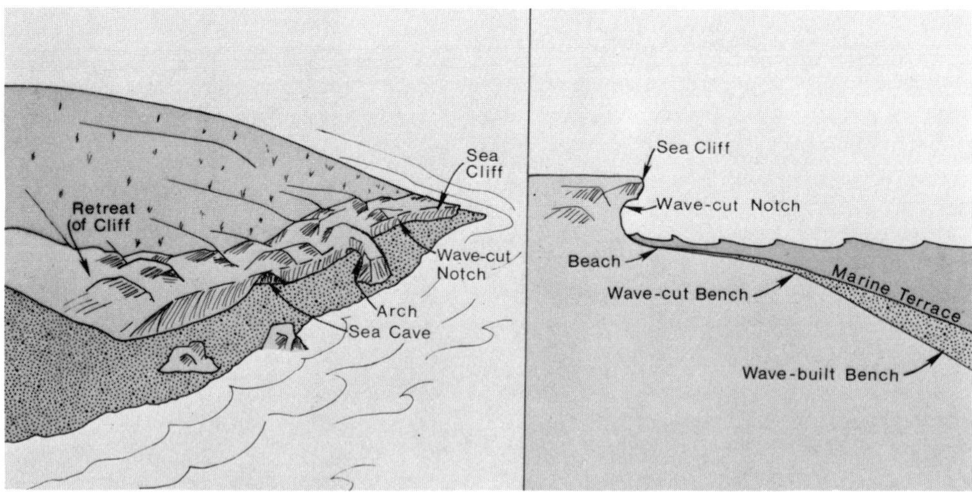

Figure 19–7

Along straight coastlines, abraded materials move along the shore in a downcurrent direction (see Figure 19–8). Where a bay or an inlet is located, the longshore currents continue straight into the bay. Material transported by longshore currents is deposited at the upcurrent edge of the bay and extends some distance across the bay mouth. This deposition feature, called a **bar** or **spit,** is usually composed primarily of sand. Sand bars can extend completely across a bay, in which case they are called *bay-mouth bars*. Tides can cut a channel through a bay-mouth bar and modify the feature into a sand bar called a *hook*.

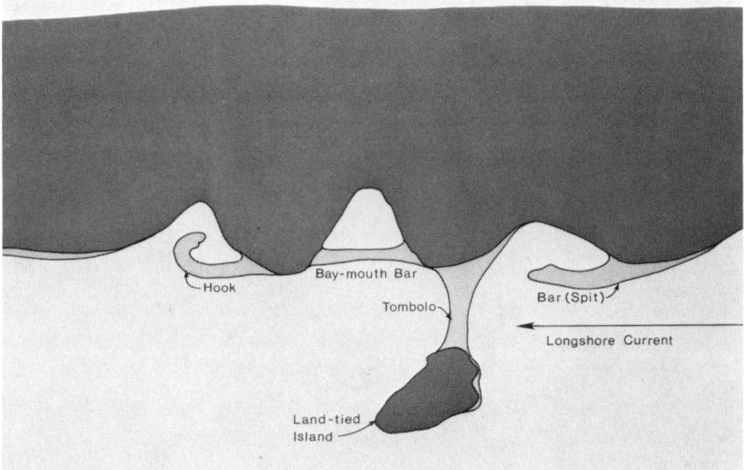

Figure 19–8

Longshore currents can pile up sand on the leeward sides of islands. In some cases, the sand may accumulate enough to connect the island to the mainland; the sand deposit is called a **tombolo,** and the island is referred to as a *land-tied island* (see Figure 19–8).

Along shallow coasts, waves strike the shore bottom and pick up sand a considerable distance from the land. As a result, breakers form a great distance offshore. The sand picked up by the breakers is deposited a short distance on the landward side of the breakers. After a period of time, the sand deposited by the breakers can accumulate to form an offshore bar. Storm waves can pile sand even higher onto the offshore bars. After the storm ceases and the sea returns to normal water level, the sand deposit dries out above sea level and forms a **barrier beach,** which is separated from the shore by a broad, shallow lagoon.

Shorelines can be classified into several types. One type, called a *submerged shoreline,* forms as portions of the land are inundated by the sea as the land bordering the sea sinks or as sea level rises (see Figure 19–9A). The actual appearance of the shoreline is dependent on the formation of the land prior to submergence. A land surface divided by stream action into valleys and hills, for example, will produce a highly irregular shoreline referred to as **ria shoreline** (see Figure 19–9B). Submerged river valleys along a ria shoreline are called **estuaries** and serve as major inlets into the coast. The ridges between stream valleys jut into the sea as *peninsulas,* and the tops of ridges may form offshore islands. Years of wave erosion tend to reduce the irregularities of a submerged coastline. Headlands are worn away, bars form across bays, and overall the coastline is straightened and shortened. Submerged coastlines are relatively common features around the world since sea levels have generally risen since the Pleistocene epoch as glaciers have melted. The Atlantic coast of the United States is a submerged coastline. One advantage of submerged coasts is their general suitability for ports and harbors.

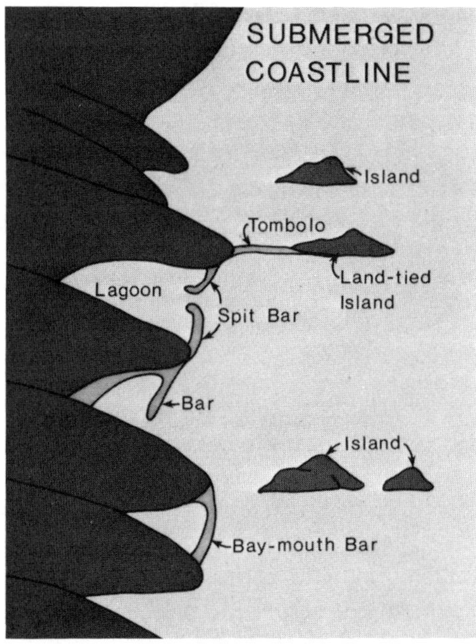

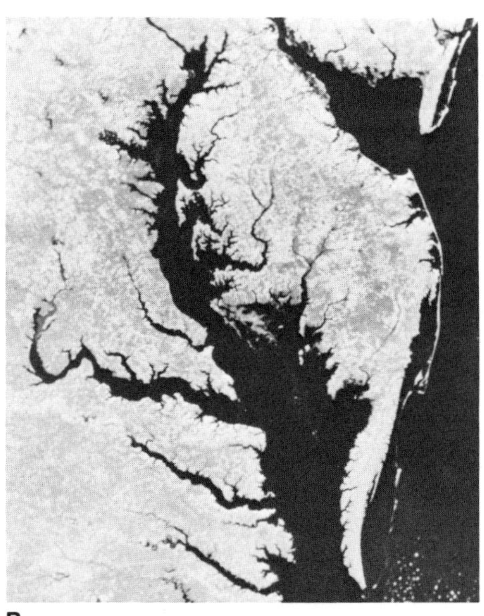

Figure 19–9

A second class of shoreline is an *emerged shoreline* (see Figure 19–10). Emerged shorelines form as the land rises relative to sea level, or as sea level declines. As a coastline emerges, wave-cut benches are exposed, and the shoreline appears very straight and gradually sloping. In many cases, due to the gradual slope of the wave-cut bench, the depth of the sea increases only

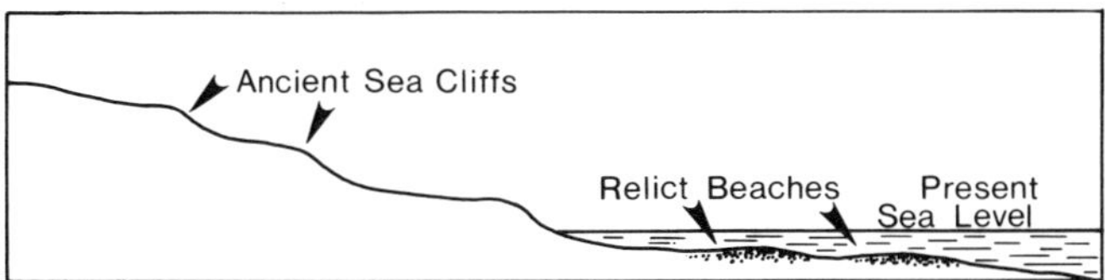

Figure 19–10

very gradually. Offshore bars and barrier beaches, characteristic of shallow water, are therefore characteristic of emerged coastlines. Emerged coastlines are usually the result of tectonic activity, which has caused the land surface to rise. The Pacific coast of the United States, for example, is an emerged coastline. In portions of the California coastline, in fact, wave-cut benches are exposed as much as 1200 feet above sea level. Ports and harbors are generally poorly developed along emerged coasts since offshore water depths are often too shallow for ocean-going vessels.

A third type of shoreline, called a *neutral shoreline,* forms where there has been no appreciable change in the elevation of the land or in sea level. Deposition is the most active process along neutral coasts. River deltas, coral reefs, and lava flows are characteristic features of neutral coasts.

Coastlines can simultaneously exhibit characteristics of submergence, emergence, and neutrality. For that reason, there is a fourth class of coastline, labeled a *compound coastline.* Features of submergence, for example, may be succeeded by emergence and neutrality, or vice versa. Compound coastlines reflect the many times sea level and land elevations have changed during the earth's history.

Questions

1. How does wind erosion compare in effectiveness with the other agents of erosion?

2. How are blowouts formed? Why is blowout formation often associated with grazing animals in arid and semiarid regions?

3. What are the various types of sand dunes? How is each type associated with wind direction and the supply of sand?

4. How does the V-shaped profile of streams in humid areas contrast with the profile of streams in arid regions? What explains this contrast?

5. What is the major difference between a pediment and an alluvial fan?

6. Why are some plateaus eroded into badlands and others eroded into mesas and buttes?

7. How do longshore currents transport objects?

8. How is a marine terrace formed?

9. What is the relationship between headlands and the deposition of eroded materials?

10. Why are submerged coastlines better suited to human uses than emerged coastlines?

Suggested Readings

BUTHER, M. 1972. Netherlands: Dutch continue to reclaim land from the sea. *Science* 176.

DAVIS, J. H. 1971. Influences of man upon coastlines. In *Man's Impact on Environment,* ed. T. R. Detwyler. New York: McGraw-Hill.

KING, C. A. M. 1972. *Beaches and coasts.* 2d ed. London: Edward Arnold.

ROWNTREE, R. A. 1974. Coastal erosion: the meaning of a natural hazard in the cultural and ecological context. In *Natural Hazards,* ed. G. White. New York: Oxford.

SHEPARD, F. P., and WANLESS, H. R. 1971. *Our changing coastlines.* New York: McGraw-Hill.

THORNBURY, W. D. 1969. *Principles of geomorphology.* 2d ed. New York: John Wiley & Sons.

Appendix

Maps

A **map** is a representation, drawn on a flat surface, of all or part of the earth's surface. Maps are valuable tools for the geographer because they make it possible to analyze large areas of the earth's surface at one time. They are used to describe and explain the distribution of both human-related features and natural phenomena on the earth. Indeed, almost any feature on the earth that can be measured and observed can be mapped.

Maps are models of the real world, but they can never accurately *duplicate* the real world because they are two-dimensional representations of a three-dimensional earth. It is physically impossible to transfer all the properties of a three-dimensional surface onto a flat surface without distortion. You have encountered this problem if you have ever attempted to flatten the skin of a peeled orange without distorting the shape of the skin. In order to maintain a close approximation of the properties of the real world, map makers, or **cartographers,** must decide which of the properties of three-dimensionality will be retained on the map and which will be sacrificed. Selecting a particular property will give the map useful qualities for some purposes. The transforming of the features from a spherical surface (earth) to a plane surface (map) is called **projection.** All maps, then, are more accurately called **map projections.** The systems of transformation of a projection are highly varied. Most important in terms of the selected system of transformation are the properties that the map will retain.

The goals of this appendix are as follows:

1. To provide you with a brief description of the basic features of a map.
2. To instruct you as to how to recognize map properties.
3. To introduce various kinds of maps.
4. To explain the reading of topographic maps.

The science of map making, called **cartography,** is a complex study, for there are many different kinds of maps, and each is compiled to satisfy a specific purpose. Many textbooks are available for those students interested in a more detailed treatment of the subject.

Properties of Maps

The basic properties that can be retained on a map are equivalency, conformality, and azimuthality. The property that is retained on a map is determined by the method in which the Geographic Grid System is projected onto the map. For example, maps drawn with the property of **equivalency,** or equal areas, retain the size of areas in the same proportion as the areas are on the earth's surface. South America, for example, is roughly eight times larger than Greenland (6,870,000 square miles versus 840,000 square miles). On an equal-area map, South America covers an area eight times larger than the area covered by Greenland. Figure A–1 is an equal-area map referred to as **Mollweide's projection.** The *proportions* of South America and Greenland are accurate, but notice what happens to the *shapes* of those areas. On equal-area maps it is necessary to distort the shapes of areas in order to retain the proper size of each region. Note that unlike meridians and parallels on the earth's surface, meridians and parallels on equal-area maps do not intersect at right angles. Consequently, although the sizes of areas can be retained, accurate shapes cannot be retained simultaneously. Thus, equal-area maps are used primarily to display the distribution of phenomena across the earth. They provide a means for making visual comparisons between the frequency of occurrence of a phenomenon in different areas. For example, the number of people per square mile or the number of bushels of corn produced per acre are just two instances of the use of an equal-area map.

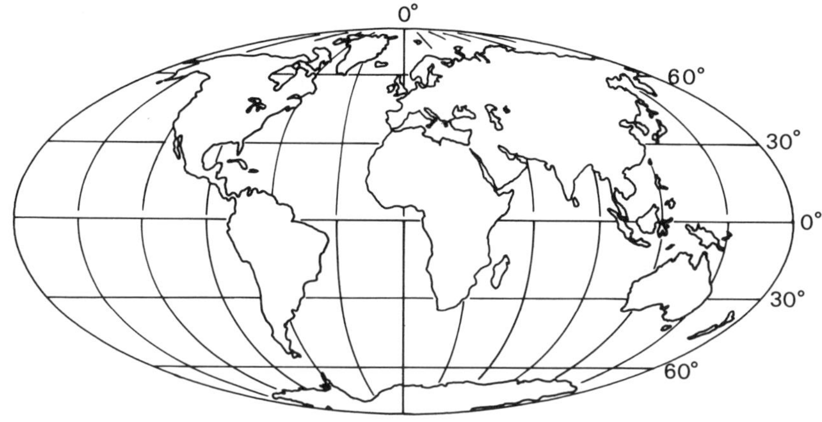

Figure A–1

Maps with the property of **conformality** retain the proper shapes of small geographical areas. Cartographers retain conformality by projecting the lines of latitude and longitude on a map such that they intersect at 90° angles just as they do on the earth's surface. Since the grid system duplicates the grid system on the earth, all compass directions are accurate on conformal maps; thus, these maps are very useful in navigating or surveying. However, when the property of conformality is retained, areal *proportions* are greatly distorted. Figure A–2 is a common conformal map projection called a **Mercator projection.** The Mercator projection has been used for navigation since the late 1500s. The areal distortion on the Mercator projection is most extreme in the polar regions where meridians remain parallel instead of converging at the poles as they do on earth. Due to the stretching apart of the meridians in such a fashion, Greenland appears almost half as large as South America—an obvious distortion of area.

Maps with the property of **azimuthality** show true compass direction from one point, or at most two points, on a map. Many azimuthal map projections also have the property of equidistance or true scale distance from one or two points. Azimuthal projections are drawn such that all direction lines **(azimuths)** drawn on the map are great circle routes. **Great circle routes** are the shortest distances between two points on a sphere. Azimuthal

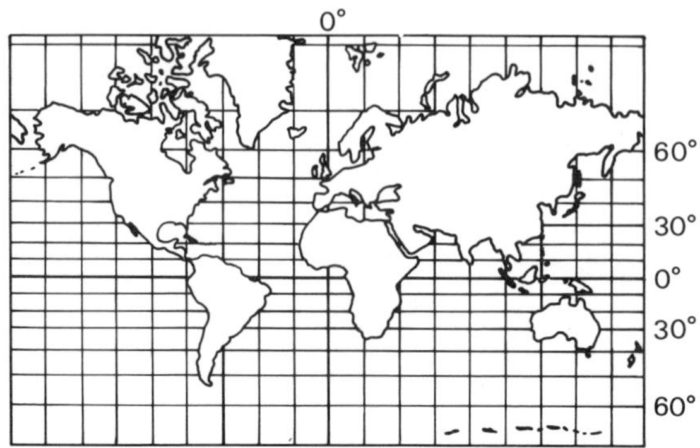

Figure A–2

projections are used as navigational aids because they plot the shortest route between two places on the earth. Figure A–3 is an example of an azimuthal-equidistant projection. The greatest degree of distortion on azimuthal projections is away from the center point of the map. Of course, by retaining the property of azimuthality on the map, cartographers sacrifice equivalency and conformality.

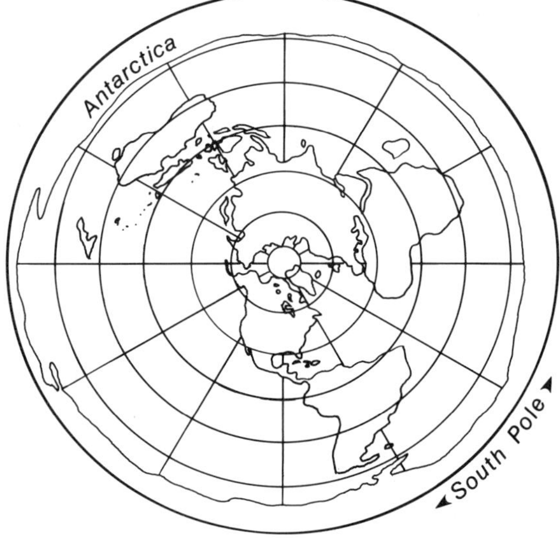

Figure A–3

Globes and Maps

It would be mentally taxing indeed to learn the properties and limitations of all the different map projections in use today. However, you should be able to look at a map and know whether it can be used to measure distances accurately from all points or whether the items that appear on the map represent an accurate distribution. Since all maps are a picture of all or part of the earth, if you first understand the properties of the earth's shape, you will be aware of the distortions that appear on a map. A globe is the most accurate representation of the earth because it retains the following features of the geographic grid from the earth's surface:

1. All parallels are parallel to one another.
2. Meridians converge at the poles.
3. Meridians are evenly spaced along any given parallel.
4. Meridians and parallels intersect at right angles.
5. Sections which are formed by the same two parallels and which extend the same longitudinal distance have the same area.
6. The scale is constant.

If any of the above features of a globe are distorted when a map is compiled, the map contains some distortion. With the exception of equivalency, none of the map properties can be retained uniformly throughout an entire projection.

You can identify equal-area projections by comparing the size of a section outlined by two parallels and two meridians on a globe with the size of the same section on the map. If the sizes of the areas are the same even though the sections may not be square on the map, the map has the property of equivalency. For example, sections A and B shown below are equal in area, but their shapes are not the same:

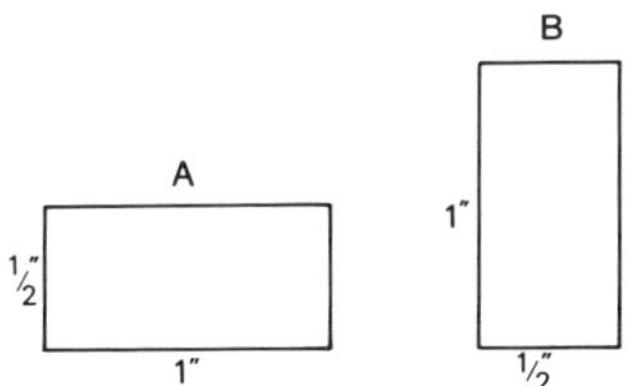

Conformality is accurate for only small sections on a map. As stated previously, meridians and parallels intersect at right angles on conformal projections and hence retain true shapes of map features. However, areas become greatly distorted. Sections A and B in the following figures, for example, conform in shape, but there is an obvious difference in their areas:

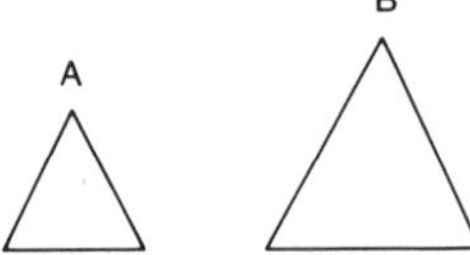

No map can have both equivalency and conformality.

Distance and direction vary greatly on maps. Distances are measured accurately only on very large-scale maps that show small areas. On small-scale maps, such as a world map, distances are usually accurate only along given parallels. Consequently, most measures of distance on small-scale maps are approximations at best. (Scale on maps will be discussed in greater detail in a later section.) Directions are accurate only over the entire map on azimuthal projections where the angles are true in all directions from any one point.

Map Essentials

All maps must contain some basic information so that they can be properly interpreted by the reader. The basic information that appears on maps is often referred to as the **map essentials:**

1. Title
2. Legend.
3. Grid system.
4. Indication of direction.
5. Scale.

Most of the map essentials are easily understood. Each map should have a **title** so that the reader knows the theme of the map. Figure A–4, for example, is entitled ''World Vegetation Types,'' a title that clearly indicates the theme of that map.

A **legend** on a map explains the colors and symbols used on the map. Without this information, the reader would not be able to decipher the information on the map. A small circle on a map could indicate anything from the place where 5000 bushels of wheat are produced annually, to the place where more than 80% of the city population consumes 100 pounds of hot dogs each year. On topographic maps, a great variety of information is shown in both colors and symbols. A legend explaining the symbols used on topographic maps is shown in Figure A–5. Of course, you cannot see the colors that usually appear in such a legend. For example, the upper left first four lines representing roads would appear in red, as would the township and section

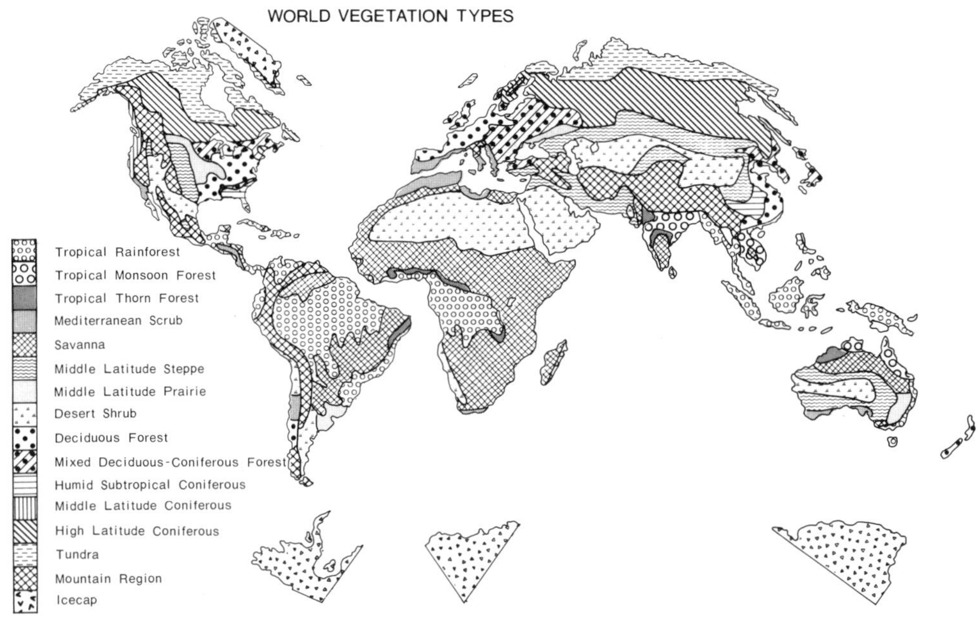

Figure A–4

lines in the right-hand half of the legend. Water symbols would appear in blue.

Maps should contain some method of determining location or, in other words, a **grid system.** Most maps contain the Geographic Grid System composed of latitude and longitude lines. This grid is labeled in various ways on a map. Maps that show small areas but great detail may include, in addition to the Geographic Grid System, a second or even third type of grid system. Topographic maps, for example, often include the U.S. Public Land Survey System and the U.S. Military Grid System. These additional grid systems are useful for more precise descriptions of phenomena on the map.

Direction on a map is generally evident whenever a grid system is employed. The Geographic Grid System is drawn such that direction relates to the geographic North Pole, or **geographic north.** However, if you use a magnetic compass to determine your direction of travel, be aware that in most places geographic north differs by several degrees from north as indicated by your compass. North as indicated by compass is termed **magnetic north.** The magnetic north pole indicates the flow of magnetism through the earth. It is not a stationary point, so its location is re-marked every five years. The magnetic north pole is at about 76°N, 100°W. Figure A–6 shows the location of magnetic north relative to geographic north. The angular distance between these two locations is referred to as **magnetic declination.** Some maps, particularly topographic maps and survey maps, show the degree of magnetic

Hard surface, heavy duty road, four or more lanes . . . . . .

Hard surface, heavy duty road, two or three lanes . . . .

Hard surface, medium duty road, four or more lanes . . . .

Hard surface, medium duty road, two or three lanes . . . . .

Improved light duty road . . . .

Unimproved dirt road and trail . . . .

Dual highway, dividing strip 25 feet or less . . . .

Dual highway, dividing strip exceeding 25 feet . . . .

Road under construction . . . .

Railroad, single track and multiple track . . . .

Railroads in juxtaposition . . . .

Narrow gage, single track and multiple track . . . .

Railroad in street and carline . . . .

Bridge, road and railroad . . . .

Drawbridge, road and railroad . . . .

Footbridge . . . .

Tunnel, road and railroad . . . .

Overpass and underpass . . . .

Important small masonry or earth dam . . . .

Dam with lock . . . .

Dam with road . . . .

Canal with lock . . . .

Buildings (dwelling, place of employment, etc.) . . . .

School, church, and cemetery . . . .

Buildings (barn, warehouse, etc.) . . . .

Power transmission line . . . .

Telephone line, pipeline, etc. (labeled as to type) . . . .

Wells other than water (labeled as to type) . . . . o Oil . . . . . o Gas

Tanks; oil, water, etc. (labeled as to type) . . . . Water

Located or landmark object; windmill . . . . o . . . .

Open pit, mine, or quarry; prospect x

Shaft and tunnel entrance . . . . Y

Horizontal and vertical control station:

    Tablet, spirit level elevation . . . . BM △ 5653

    Other recoverable mark, spirit level elevation . . . . △ 5455

Horizontal control station: tablet, vertical angle elevation VABM △ 9519

    Any recoverable mark, vertical angle or checked elevation △3775

Vertical control station: tablet, spirit level elevation . . . . . BM ✕ 957

    Other recoverable mark, spirit level elevation . . . . ✕ 954

Checked spot elevation . . . . ✕ 4675

Unchecked spot elevation and water elevation . . . . ✕ 5657 870

Figure A–5

324

Boundary, national................................. —— — — ——

    State.............................. —— — — —

    County, parish, municipio........................ —— — — —

    Civil township, precinct, town, barrio............... — — — —

    Incorporated city, village, town, hamlet............. —–————

    Reservation, national or state.................... —— . ——— .

    Small park, cemetery, airport, etc................. ..——————

    Land grant..................................... —— .. ——— ..

Township or range line, United States land survey...... ————————

Township or range line, approximate location.......... ——————-

Section line, United States land survey................ ————————

Section line, approximate location.................... —— — — —

Township line, not United States land survey ....

Section line, not United States land survey ....

Section corner, found and indicated .... +.........+

Boundary monument: land grant and other............ □.........□

United States mineral or location monument ....   ▲

Index contour	Intermediate contour..
Supplementary contour	Depression contours ..
Fill...............	Cut..............
Levee.............	Levee with road......
Mine dump..........	Wash..............
Tailings...........	Tailings pond.......
Strip mine..........	Distorted surface.....
Sand area..........	Gravel beach.......

Perennial streams	Intermittent streams..
Elevated aqueduct....	Aqueduct tunnel......
Water well and spring.. o ... o	Disappearing stream..
Small rapids.........	Small falls ....
Large rapids........	Large falls ....
Intermittent lake.....	Dry lake............
Foreshore flat.......	Rock or coral reef....
Sounding, depth curve.	Piling or dolphin.....
Exposed wreck.......	Sunken wreck.......
Rock, bare or awash; dangerous to navigation..........	

Marsh (swamp).......	Submerged marsh ....
Wooded marsh.......	Mangrove ....
Woods or brushwood..	Orchard............
Vineyard...........	Scrub.............
Inundation area......	Urban area ....

Figure A–5, Continued

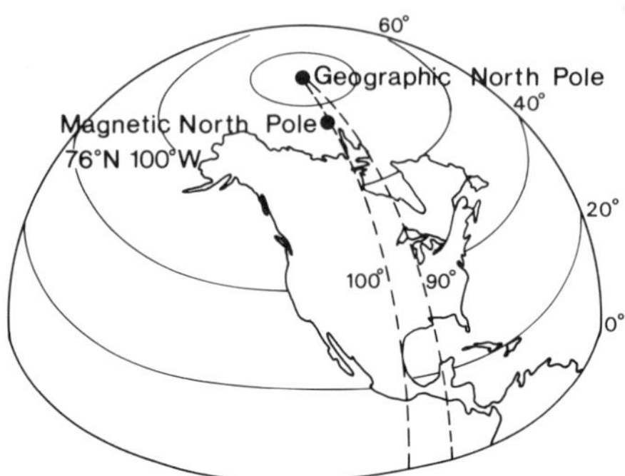

Figure A–6

declination with arrows at the base of the map. On Figure A–5 the magnetic declination is 18°.

Scale on a map is the ratio between a unit of distance on the map and the distance on the earth's surface measured in the same units. One inch or one centimeter on the map will equal so many inches or centimeters on the earth's surface. Scale can be shown in three ways on a map:

1. **Representative fraction (RF).** The map scale is written in ratio or fractional form, such as 1 : 24,000 or $\frac{1}{24,000}$. In this example, one unit on the map equals 24,000 of the same unit on the earth's surface.

2. **Graphic scale.** Graphic scale on a map is a line which is drawn on the map and which is subdivided into units of distance. You can use the graphic scale by placing a straight edge between two places on the map, marking the distance between the two places, and then measuring this distance directly on the scale shown on the map. Typical graphic scales from a topographic map are shown below:

1	1/2	0	1	2	3	4 Miles

3000	0	3000	6000	9000	12000	15000	18000	21000 Feet

1	.5	0	1	2	3	4	5 Kilometers

3.   **Verbal scale.** A verbal scale is a written message on the map
     that states the scale to the reader. An example of a verbal
     scale would be the statement, ''One inch equals one mile.''

Maps can be drawn to any scale, but several scales are more commonly
applied than others. Among the more commom are those given in Table A–1.

Table A–1

RF	One inch on the map equals:
1 : 24,000	2000 feet
1 : 62,500	0.986 mile (about 1 mile)
1 : 100,000	1.58 miles
1 : 1,000,000	15.78 miles
1 : 2,500,000	40 miles (approximate)
1 : 75,000,000	1200 miles (approximate)

Maps are often referred to as **large-scale** or **small-scale** maps.
Large-scale maps are those whose representative fraction is larger than
1 : 1,000,000. Topographic maps, for example, are drawn at the scale of
either 1 : 24,000 or 1 : 62,500. State road maps are drawn at large scale, such
as 1 : 82,500. Large-scale maps show great detail but cover limited areas.
Small-scale maps have representative fractions that are smaller than approximately 1 : 1,000,000. They show extensive areas but little detail, such as an
entire continent or the whole world (see Figure A–4). Many wall maps of the
world are drawn at a scale of 1 : 2,500,000, and world maps in an atlas are
often at the scale of 1 : 75,000,000. Small-scale maps provide only a very
generalized picture of the earth's surface.

Topographic Maps

Topographic maps are large-scale maps that show both the horizontal and the
vertical dimensions of a given area; thus, they are useful tools for interpreting
the landscape. **Topography** can be defined as the physical features of a
region. **Topographic maps** are representations of the physical features of a
region drawn to scale on a flat surface. In addition, topographic maps display
many of the cultural features of the region by use of map symbols (see Figure
A–5). The representation of both physical and cultural features of a region
makes topographic maps particularly useful to the geographer, who attempts

to explain the location and distribution of those features within a spatial context.

The horizontal pattern of physical and cultural features across the landscape is only one dimension of topographic maps. The **relief** of the landscape, or the variations in elevation of the earth's surface, are also shown on topographic maps by the application of isopleths. **Isopleths** are lines on a map that connect points of equal value. Isopleths that connect points of equal elevation are called **contour lines** or **contours.** All places along the same contour line are at the same elevation.

Figure A–7 illustrates contour lines. If you were to mark a trail around a small circular hill surrounded by a flat plain, always making certain that you stayed exactly 10 feet above the flat plain, your trail would appear as in Figure A–7. If you repeated the process of marking a trail around the hill several times, but increased your elevation by 10 feet each time, the hill would be marked as shown in Figure A–8. An airplane flying at high altitude over the marked hill would see a pattern as shown in Figure A–9. This is the type of pattern that appears on a topographic map.

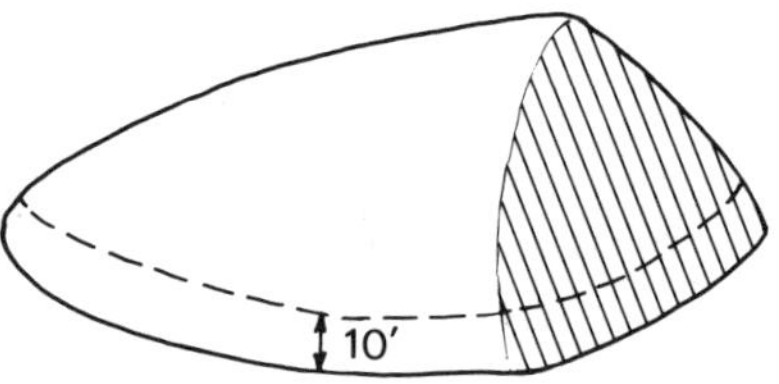

Figure A–7

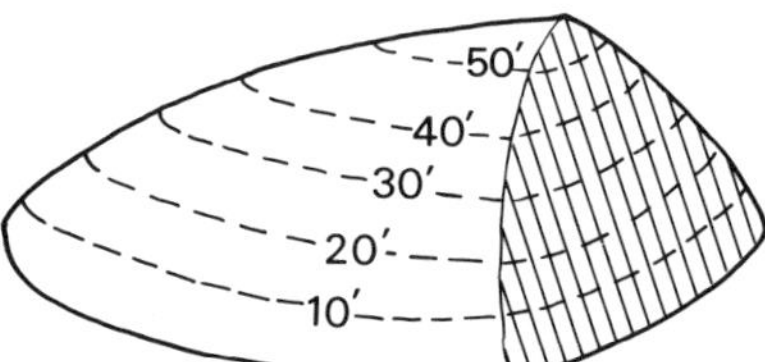

Figure A–8

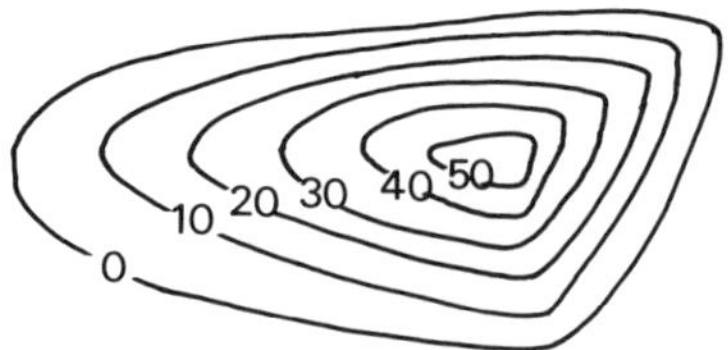

Figure A–9

From Figures A–7, A–8, and A–9, we can list two basic characteristics of contour lines:

1.   They are continuous lines.
2.   They never cross one another.

You may have noticed one other feature of contour lines from Figure A–9: The closer the spaces between contours, the steeper is the slope of the land.

On a topographic map the contours are labeled at various intervals, depending on the number of feet that the elevation changes between contours. On most topographic maps of regions where slopes are gradual to moderate, contours are spaced 10 or 20 feet apart. Each fifth contour is then labeled and drawn in a heavier line on the map. In regions where slopes are very steep, a larger change in elevation between contours is necessary, such as 40, 50, or even 100 feet. The elevation change between contours on a map is referred to as the **contour interval.** The contour interval can be found at the base of each topographic map. The contour interval should be checked each time you pick up a topographic map.

In most regions of the world, the landscape has been dissected to some degree by streams that flow in channels. When contour lines meet a stream channel, they bend into a V-shape, with the point of the V directed upstream. In other areas, depressions form on the land. Depressions are shown by contours that have **tick marks** along them. Where tick marks appear along a contour, elevations decrease at the contour interval on the map.

Acknowledgments

Media

Frame **14–10** painting by Heinrich Berann, courtesy of the Aluminum Corporation of America.

Frames **5–34, 5–40,** and **8–26** from Richard A. Anthes et al., *The Atmosphere,* 2d ed. (Columbus, OH: Charles E. Merrill, 1978).

Frames **4–22, 4–24, 4–36, 5–1, 6–8B,** and **6–13** from Richard A. Anthes, *Weather Around Us* (Columbus: Merrill, 1976).

Frames **9–8B, 9–27A, 9–28A, 10–31, 11–15,** and **12–8** courtesy of M. Beatty.

Frame **6–2** courtesy of Peter Black.

Frames **8–47, 9–17B, 13–1, 13–15, 13–17B, 13–24B, 15–1B, 15–11, 15–13B, 16–4, 16–8, 16–14, 17–17, 18–10, 18–37B, 18–41, 19–1A,** and **19–7** courtesy of George D. Brown, Jr.

Frames **13–3, 13–16, 13–17A, 13–24A, 14–3, 14–7, 14–9, 14–15, 14–21, 14–22, 15–3B, 15–8A, 15–25,** and **17–14** from George D. Brown, Jr., and George T. Ladd, *The Earth: Man's Geologic Environment* (Columbus: Merrill, 1976).

Frame **16–17** from George D. Brown, Jr., and George T. Ladd, *Excursions in Geology.* Copyright © 1973 by Houghton Mifflin Company.

Frames **18–6B, 18–27B, 18–35,** and **18–36** courtesy of A. Cahow.

Frame **15–12** courtesy of Consulate General of Iceland.

Frame **13–8** by Steven Cremer.

Frame **10–15B** courtesy of John Deere Company.

Frame **5–32** courtesy of John Dietrich, ESSA.

Frames **14–16** and **14–18** from Robert S. Dietz and John C. Holden, *Journal of Geophysical Research* 75, pp. 4943–44, 4947, and 4949, 1970, copyrighted by American Geophysical Union.

Frames **14–19** and **14–20** from "The Breakup of Pangaea" by Robert S. Dietz and John C. Holden. Copyright © by Scientific American, Inc. All rights reserved.

Frames **8–20B, 8–58, 9–30A, 10–12, 10–14A** and **B, 12–10, 12–14, 12–15, 12–16,** and **12–34** courtesy of Educational Filmstrips, Huntsville, Texas.

Frame **17–16B** by ERTS.

Frame **11–22** courtesy of Farmers and Manufacturers Beet Sugar Association.

Frame **8–26** by Alistair Fraser; © Alistair Fraser.

Frame **6–20** by T. Fujita.

Frame **6–15:** NOAA/NSSL photo courtesy of Dr. J.H. Golden.

Frame **6–21:** Waterspout north of Key West, Florida, Sept. 10, 1969. Photo courtesy of Dr. J.H. Golden.

Frame **8–49** courtesy of R. Gonowski.

Frames **4–8, 5–3, 19–25, 19–27, 19–36B,** and **19–37** from M. Grant Gross, *The Ocean World* (Columbus: Merrill, 1976).

Frames **15–1A, 15–3A, 15–8B,** and **15–13A** courtesy of Hawaii Visitors Bureau.

Frame **15–14** by J.C. Hepburn.

Frames **4–38B, 6–15,** and **8–37** courtesy of Charles S. Hosler.

Frames **1–12B, 4–37, 10–15A, 11–14, 12–32, 13–4A** and **B, 13–9, 13–10A** and **B, 13–11, 13–13, 13–14, 13–20, 13–21A** and **B, 13–25A** and **B, 15–19B, 15–21, 15–22, 15–24A** and **B, 15–26, 15–27, 15–35, 16–4, 16–5A, 16–7B, 16–9, 16–12A** and **B, 16–15, 16–20, 16–22, 16–24B, 16–25B, 17–2, 17–9A, 17–12, 17–20, 17–23, 17–25, 18–8A** and **B, 18–9, 18–11, 18–14A, 18–15, 18–16, 18–24B, 18–27A, 18–32, 18–34, 19–5A** and **B, 19–13, 19–15, 19–16,** and **19–30** courtesy of R. Janke.

Frame **6–5** courtesy of C. Jenner.

Frame **6–14** by Gary Ladd.

Frame **16–19** courtesy of Joseph Magnuson.

Frame **4–1** by Betty Ann Morse.

Frames **1–1, 2–7, 17–29, 19–34B,** and **19–36A** courtesy of NASA.

Frame **16–8** courtesy of National Park Concessions, Inc.

Frames **15–4B** and **15–10** by W.L. Keller, courtesy of National Park Service.

Frame **19–22** courtesy of National Park Service.

Frame **6–9** courtesy of National Weather Service.

Frames **3–21B, 6–27A, 8–53,** and **18–14B** courtesy of Naval Photographic Center.

Frames **11–12** and **12–2A** courtesy of NCGE Slide Library.

Frame **18–19** courtesy of S. Ness.

Frames **4–34, 4–37, 6–1, 6–19, 6–27B, 6–28,** and **14–1** courtesy of NOAA.

Frame **11–2** from Eugene P. Odum, *Fundamentals of Ecology,* 3d ed. (Philadelphia: Saunders, 1971).

Frame **16–13** reproduced by permission from the ANNALS of the Association of American Geographers, Volume 40, 1950, L.C. Peltier. Adapted in W.D. Thornbury, *Principles of Geomorphology,* 2d ed. (New York: John Wiley & Sons, 1969).

Frame **6–22** by Robert Sheets, National Hurricane Research Laboratory.

Frames **8–17, 18–39, 18–40,** and **19–19** by John S. Shelton.

Frames **10–6, 10–9, 10–27,** and **10–35** courtesy of Soil Science Society of America.

Frames **3–18B, 4–6, 4–19, 4–43B, 5–7B, 7–23A, 8–3, 8–14A** and **B, 8–54, 10–32, 11–11A, 11–18, 12–3, 12–12, 15–5,** and **15–33B** courtesy of A.R. deSouza.

Frame **18–17B** courtesy of Swiss National Tourist Office.

Frames **9–13B** and **19–26A** and **B** courtesy of G. Tanner.

Frame **7–10** courtesy of U.S. Bureau of Reclamation.

Frames **8–45, 10–18A, 10–29B, 11–3, 11–9, 12–17, 12–19, 12–21,** and **12–27** courtesy of U.S. Forest Service.

Frames **19–6A** and **B** courtesy of U.S. Soil Conservation Service.

Frames **7–16, 8–42A, 9–5B, 10–11, 10–36, 16–16, 17–1, 17–5B, 17–11, 17–13,** and **17–20** courtesy of University of Wisconsin Extension Service.

Frames **4–43A, 8–7, 8–16, 8–30B, 10–33B, 11–16, 12–4, 12–7,** and **12–9** courtesy of I. Voegler.

Frame **7–17** courtesy of Water Information Center, Inc., from *Water Atlas of the United States,* 3d ed., 1973, plate 23.

Frames **3–19B, 3–44A, 12–6, 12–30, 15–2B, 15–6, 15–7, 19–1B,** and **19–4** courtesy of R. Willis.

Frames **5–7A, 6–12, 10–18B, 10–20, 11–4, 12–23,** and **16–21B** courtesy of Wisconsin Department of Natural Resources.

Frames **3–20B, 7–28, 9–5A, 9–7, 9–8A, 9–11B, 9–13A, 9–19B, 9–25, 9–26, 9–27B, 10–11, 10–22, 10–24, 10–34, 11–15A** and **B,** and **11–19** courtesy of Wisconsin Soil Conservation Service.

Book

Figure **14–4** painting by Heinrich Berann, courtesy of the Aluminum Corporation of America.

Figures **13–4** and **16–3** courtesy of George D. Brown, Jr.

Figures **14–6, 14–8, 14–12, 17–3, 17–4,** and **19–2** from George D. Brown, Jr., and George T. Ladd, *The Earth: Man's Geologic Environment* (Columbus: Merrill, 1976).

Figure **18–2B** courtesy of A. Cahow.

Figures **13–3** and **13–5** by Steven Cremer.

Figure **14–10** from Robert S. Dietz and John C. Holden, *Journal of Geophysical Research* 75, pp. 4943–44, 4947, and 4949, 1970, copyrighted by American Geophysical Union.

Figure **17–5** by ERTS.

Figure **8–3** reprinted by permission, from S.R. Eyre, *Vegetation and Soils* (Chicago: Aldine Publishing Company); © S.R. Eyre 1968.

Figures **14–4** and **19–5** from M. Grant Gross, *The Ocean World* (Columbus: Merrill, 1976).

Figures **13–1, 16–1A, 16–5B, 17–2A, 18–3, 18–9,** and **19–1** courtesy of R. Janke.

Figures **17–4** and **19–9** courtesy of NASA.

Figure **15–3** by W.L. Keller, courtesy of National Park Service.

Figure **11–1** from Eugene P. Odum, *Fundamentals of Ecology,* 3d ed. (Philadelphia: Saunders, 1971).

Figure **15–11** courtesy of James L. Ruhle and Associates.

Figures **12–2, 12–3,** and **12–10B** courtesy of A.R. deSouza.

Figures **12–4, 12–5,** and **12–6** courtesy of U.S. Forest Service.

Figure **12–10A** courtesy of I. Voegler.

Figure **15–2** courtesy of R. Willis.

Figure **16–4B** courtesy of Wisconsin Department of Natural Resources.

Glossary

Aa lava   Lava of high silica content that solidifies into an irregular, jagged mass.

Abrasion   The mechanical process of erosion by rock particles transported by running water, moving ice, wind, and waves.

Absolute humidity   The mass of water vapor per unit volume of air, expressed in grams per cubic meter or in grains per cubic foot.

Adiabatic heating and cooling   Changes in temperature due to the expansion or compression of air without the addition or withdrawal of outside heat.

Advection   The movement of air in a horizontal direction.

Aeration zone   The top layer of soil moisture above the water table.

Air mass   An extensive body of air within which temperature and moisture conditions are essentially uniform along a horizontal plane.

Albedo   The radiation that is lost to outer space by direct reflection.

Alfisol   One of the ten orders of the U.S. Comprehensive Soil Classification. Alfisols possess a light-colored surface horizon and a distinct clay layer.

Alluvial fan   A fan-shaped alluvial deposit at the foot of a mountain stream or at another place where the velocity suddenly decreases.

Alluvium   Unconsolidated particles of gravel, sand, silt, and clay deposited by running water.

Alpine glacier   A glacier that forms in mountains.

Anticline   A rock upfold where the strata dip downward away from the center of the fold.

Anticyclone   An area of relatively high atmospheric pressure.

Aphelion   The place in the orbit of a planet where the planet is furthest from the sun. The earth's aphelion occurs on July 4.

Aquifer   A layer of permeable rock that can hold water.

Arête   A narrow, jagged rock ridge between two cirques.

Aridisol   One of the ten orders of the U.S. Comprehensive Soil Classification. Aridisols have shallow horizons, soluble minerals, and coarse textures, and they are common in desert regions.

Artesian well   A type of well where water pressure in a confined aquifer forces a continuous flow of water to the surface.

Asthenosphere   The upper layer of the mantle between a depth of 50 and 450 miles.
Avalanche   A large mass of snow or ice that moves rapidly down a mountain slope.

Bar   A ridge of mud, sand, or gravel deposited by water currents, especially at the mouth of a river or parallel to and a short distance from a beach.
Barometer   An instrument for measuring atmospheric pressure.
Barrier beach   A sandy bar parallel to a coastline and separated from it by a lagoon.
Base flow   The amount of water supplied to a surface stream by groundwater.
Base level   The lowest level to which a stream can erode its channel; normally assumed to be sea level.
Batholith   A large mass of intrusive igneous rock that has no apparent floor and covers an area greater than 40 square miles.
Bed load   Unconsolidated material carried along the bed of a stream channel without being carried in suspension.
Bergeron process   A process of raindrop formation that depends upon freezing nuclei to attract cloud droplets.
Biosphere   The zone of life at or near the earth's surface.
Blowout   A hollow or depression formed by the removal of loose particles by wind; most common in arid and semiarid regions.
Boreal forest   Coniferous forest in North America located at high latitude.
Boulder train   A series of erratics deposited by a glacier.
Broad-leaved   Having relatively broad, flat leaves as opposed to narrow leaves or needles.
Butte   A small, but prominent, flat-topped hill remaining after the dissection of a plateau by erosion.

Calcification   A soil-forming process in arid and semiarid climates; characterized by the accumulation of calcium carbonate in the B horizon.
Caldera   An enlarged volcanic crater that forms from the removal of the top of a volcano by eruption.
Calving   The breaking away of an iceberg from an ice mass into the sea.
Canopy   An almost continuous layer of vegetation formed at the crowns of tall trees, especially as in the tropical rainforest.
Capillary action   The movement of soil water toward the soil surface due to evaporation.
Carbonation   A form of chemical weathering of rocks by rainwater containing carbon dioxide in solution.
Cavitation   A form of stream erosion in turbulent water as air bubbles collapse upon collision with the stream channel.
Cementation   A form of lithification whereby the pore spaces between sediment particles are filled with a cementing agent, usually a dissolved mineral, which hardens into a rock.
Chapparal   A type of evergreen scrub vegetation found in California and northwestern Mexico.
Chemical weathering   A form of weathering whereby rocks break apart due to chemical processes such as oxidation, carbonation, and hydrolysis.
Chinook   A dry, warm foehn wind that blows down the east slope of the Rockies.
Cirque   A large, amphitheater-like rock basin at the head of a glaciated valley near the summit of a mountain.
Clastic rock   A sedimentary rock consisting of the broken particles of other rocks. Conglomerate is a clastic sedimentary rock.
Cleavage   The tendency of some rocks to break along definite lines.

Climate   An average statement of the weather conditions of a place or region over an extended period of time.

Climax vegetation   The final form of natural vegetation of a long-term plant succession; established in a physical environment that is in a stage of equilibrium.

Climograph   A graph on which the temperature and precipitation data of a place are plotted.

Col   A notch or pass through a ridge or mountain range due to the erosion of back-to-back cirques.

Cold front   The boundary between a mass of warm air and an advancing mass of colder air.

Collision-coalescence process   A process of raindrop formation whereby cloud droplets of varying size collide as they circulate within a cloud, combine into larger cloud droplets, and eventually grow into a raindrop.

Compaction   A form of lithification whereby unconsolidated particles are consolidated due to the weight of overlying materials.

Concordant   Lying parallel to the general lines of the rock strata.

Condensation   The physical process of converting water vapor into the liquid stage.

Condensation level   The elevation in the atmosphere at which air reaches the dew point.

Condensation nuclei   Particles of dust, salts, and microorganisms in the atmosphere upon which water vapor condenses.

Conduction   A process of heat transfer from warm objects to cooler objects through direct contact.

Conifer   A cone-bearing plant (usually a needleleaf tree).

Continental glacier   The largest of glaciers. A continental glacier, also referred to as an *ice sheet,* covers all or a large portion of a continent.

Continental shelf   The gently sloping margins of a continent submerged beneath the sea; extends from the shore to a place where ocean depth reaches from 350 to 600 feet.

Continental shield   A continental block of the crust that has been relatively stable over a long period of time and has undergone only gentle folding.

Convection   A localized, vertical motion in the air involving a heat transfer away from the source of heat.

Cooling degree day   The difference in degrees in 24 hours between a base temperature of 70°F and an average daily temperature above 70°F.

Cores of the earth   The innermost layers of the earth. The outer core is liquid, and the inner core is solid.

Coriolis force   The deflecting force of the earth's rotation on all free-moving bodies in the atmosphere.

Crater   A steep-sided pit or depression at the peak of a volcano.

Creep   The imperceptible movement of soil and rock downslope due to gravity.

Crevasse   A crack in the surface of a glacier caused by a strain on the glacial ice.

Crust of the earth   The outermost layer of the earth's surface above the Mohorovičić discontinuity.

Crystallization   The process of crystal formation; common in intrusive igneous rocks.

Cyclone   An area or region of low barometric pressure; also the circulation of air about a low-pressure cell.

Deciduous tree   A tree that sheds its leaves at least once during a year.

Deflation   The removal of loose particles from the earth's surface by the wind.

Degradation   The general lowering of the land surface by erosion.

Delta   An alluvial deposit formed in standing water at or near the mouth of a stream.

Deposition   The laying down of material that has been transported by running water, wind, and ice.

Desert pavement (reg) An area of desert covered by a thin layer of wind-polished pebbles.

Dessication The processes of drying out or losing water.

Dew point The temperature to which a parcel of air must be cooled in order for saturation (100% relative humidity) to occur.

Diastrophism The processes of massive earth movement and rock deformation, except vulcanism.

Dike A tabular intrusive igneous structure that lies across the structure of the surrounding rock.

Direct lifting The erosion of a stream channel as rock particles are lifted free of the channel by the action of overturning water.

Discordant Cutting across the general lines of the surrounding structure.

Distributary An individual channel that splits from the main channel and does not rejoin the main stream.

Drift All materials deposited by glacial activity.

Drumlin A smooth hill of glacial drift, deposited directly by the ice, with its long axis parallel to the direction of ice movement.

Earthflow A mass of saturated material that flows downslope under the influence of gravity.

Earthquake Vibrations of the earth caused by sudden internal movements in the earth.

Ecliptic The path described by the earth around the sun in the course of a year.

Effluent stream A stream whose channel lies below the water table level and which is fed by groundwater.

Eluviation The removal of soil material in suspension or solution from a layer or layers of the soil.

Emerged coastline A shoreline that has risen relative to sea level so that areas formerly under water have become dry land.

Entisol One of the ten orders of the U.S. Comprehensive Soil Classification. Entisols do not have developed horizons or profiles.

Eolian Wind-eroded or wind-deposited.

Ephemeral stream A stream or portion of a stream that flows only when it is raining.

Epiphyte A plant that grows on another plant and derives its nourishment directly from the air.

Equinox One of the two days each year when the vertical rays of the sun strikes the earth exactly on the equator, resulting in equal day and night at all latitudes.

Erg A sandy desert, especially the Sahara Desert.

Erosion The wearing away of land surfaces by moving water, ice, or wind.

Erratic A stone or boulder carried by ice to a location where the bedrock differs in composition.

Esker A meandering ridge of stream-deposited glacial drift formed by streams in ice tunnels beneath a glacier.

Estuary An inlet or arm of the sea, especially where the mouth of a river is invaded by the sea.

Evaporation The process by which a liquid is converted into a gaseous state.

Exfoliation The breaking away of shell-like layers of rocks from a main rock mass.

Extrusive igneous rocks Rocks or features that form while exposed to the atmosphere on the earth's surface.

Fault A crack in bedrock along which movement has occurred.

Field capacity The percentage of water remaining in a soil two or three days after the soil was saturated.

Fiord A glaciated valley along a coastline that has filled with seawater.

Firn (névé)   Partially compacted, granular snow that is in an intermediate stage between fresh snow and glacial ice.

Floodplain   An area adjacent to a stream channel that is subject to flooding.

Foehn   A warm, dry wind that descends leeward mountain slopes.

Folding   The bending of rock layers as a result of compressive forces in the earth's crust.

Foliation   A feature of metamorphic rocks whereby minerals line up in parallel bands.

Footwall block   The rock mass beneath the fault line.

Geoid   The exact shape of the earth.

Geomorphic cycle   A model of landform evolution developed by W.M. Davis.

Geostrophic circulation   The direction of air motion when the pressure gradient force, directed toward low pressure, is directly opposite and equal to the Coriolis force.

Geyser   A type of thermal spring that intermittently ejects water with considerable force.

Graben   A valley formed by the sinking of a block of the earth's crust between two nearly parallel faults.

Groundwater   Subsurface water below the water table.

Hamada   A rock desert consisting of a bare, wind-scoured surface.

Hanging valley   A tributary valley in a glaciated mountain region whose floor is considerably higher than the floor of the main valley into which it leads.

Hanging wall block   The rock mass above a fault.

Headland   Any projection of the land into the sea; generally, a narrow and steep-sided ridge.

Heating degree day   The difference in degrees between the average daily temperature and a maximum base temperature of 65°F.

Histosol   One of the ten orders of the U.S. Comprehensive Soil Classification. Histosols are characterized by a high organic content.

Horn   A sharp-featured mountain peak formed when two or more cirques erode the same rock mass from different directions.

Horst   A section of the earth's crust uplifted between two faults to form an upland.

Humidity   Water vapor in the air.

Humus   The decomposed remains of plants in the soil.

Hurricane   An intense tropical storm with wind speeds of at least 75 mi/hr.

Hydrologic cycle   The endless interchange of water between the air, land, and sea.

Hydrology   The scientific study of the waters of the earth.

Hydrolysis   A form of chemical weathering whereby water forms a permanent chemical union with a mineral.

Hydrophyte   A plant that lives in water or saturated soil.

Iceberg   A mass of land ice that has broken away from its parent formation and has moved out to sea.

Illuviation   The removal of soil material from one soil horizon and its subsequent deposition in another horizon.

Inceptisol   One of the ten orders of the U.S. Comprehensive Soil Classification. Inceptisols are characterized by poor profile development and form under greatly varying environments.

Inclination   The tilt of the earth's axis 23½° from the perpendicular.

Influent stream   A stream whose channel lies above the water table.

Inselberg   A prominent rock mass rising above a mean level erosion surface in an arid region.

Insolation The solar energy that reaches the surface of the earth.

Interfluve The area of land between two rivers.

Intrusive igneous rocks Rocks or features that form beneath the earth's surface.

Inversion An atmospheric condition in which temperature increases with height.

Ionosphere The portion of atmosphere above the stratosphere; that is, the mesosphere, and thermosphere.

Island arc A chain of islands formed by volcanic activity along the margin of an oceanic trench.

Isobar A line on a map that connects points of equal pressure values.

Isotherm A line on a map that connects points of equal temperature values.

I.T.C.Z. **Intertropical convergence zone.** A shifting trough of low pressure in tropical latitudes associated with intense solar heating. The location of the I.T.C.Z. is characterized by abundant precipitation.

Jet stream A meandering, narrow band of high-altitude winds of exceptional velocity.

Joint A crack in a rock mass in a place of weakness.

Jungle A type of tropical forest growth typified by a very dense undergrowth.

Kame A roughly cone-shaped hill of glacial drift deposited by meltwater in contact with glacial ice.

Katabatic wind A cold wind that blows downhill due to gravity.

Kettle A circular hole in glacial drift caused by the burial of an ice mass in drift which subsequently melted.

Laccolith An intrusive igneous feature that forces up the overlying rock layers into a blister-like form.

Lacustrine plain The floor of a former lake filled in by deposition; generally associated with continental glaciers.

Landslide A general term for motion of earth materials downslope due to gravity.

Lapse rate The rate of temperature change in the atmosphere with vertical air motion.

Latent heat The amount of heat energy necessary to change the state of a substance; for example, water vapor to water.

Laterization The process of soil development in hot, humid regions where soluble minerals are removed from the soil, resulting in an abundance of iron and aluminum.

Latitude The angular distance of a place north or south of the equator.

Lava Molten rock on the surface of the earth.

Leaching The removal of materials in solution by the movement of water down through the soil.

Liana A woody, climbing vine common in tropical rainforests.

Lightning A visible electrical discharge in the atmosphere.

Lithification The conversion of accumulated sediments into rock through cementation, compaction, and dessication.

Lithosphere The solid portion of the earth's crust.

Loam A soil containing a mixture of the different soil texture particles.

Loess Unconsolidated layer of silt and clay deposited by wind action.

Longitude The angular distance of a place east or west of the prime meridian.

Longshore current The movement of water along a shoreline as a result of waves breaking at an angle to the shore.

Macronutrient   A chemical element necessary in large amounts for the growth of plants; usually applied artificially in fertilizers.

Magma   Molten rock beneath the earth's surface.

Mantle   The middle layer of the earth between the crust and the core.

Maquis   A low scrub of evergreen vegetation found in Mediterranean areas.

Mass movement   A general term for the movement of earth materials downslope due to the pull of gravity.

Meander   A loop in a river as it winds back and forth across a stream valley.

Meander scar   A meander that has been abandoned by a stream and filled in by deposition and vegetation.

Mechanical weathering   The physical disintegration of rock into smaller pieces by agents of the weather; no chemical alteration is involved.

Meridian   A line of longitude.

Mesa   A flat-topped, steep-sided erosional remnant that has a greater horizontal than vertical dimension.

Mesophyte   A plant with moderate water needs. Most trees are mesophytes.

Mesosphere   A layer of the atmosphere between the stratosphere and the thermosphere.

Mesothermal climates   A group of world climates with moderate temperatures and rainfall and ample sunshine; that is, humid subtropical, marine west coast, and Mediterranean climates.

Metamorphic   One of the three basic rock types; formed from other rocks due to changes in temperature and pressure.

Micronutrient   A chemical element necessary in soils in only small amounts for the growth of plants.

Microthermal climates   A group of world climates generally characterized by moderate precipitation, warm summers, and long, cold winters.

Mid-ocean ridge   A long, continuous, underwater mountain range that lies roughly parallel to continental margins; formed by volcanic outpourings from the asthenosphere.

Millibar   A unit for measuring atmospheric pressure; equal to 1000 dynes per square centimeter. 33.85 millibars = 1 inch of mercury.

Mineral   An inorganic substance composed of two or more chemical elements.

Mirage   An optical illusion caused as light is refracted as it passes through a layer of warm surface air to a layer of cooler air above.

Mohorovičić discontinuity (Moho)   The boundary between the earth's crust and its mantle.

Mollisol   One of the ten orders of the U.S. Comprehensive Soil Classification. Mollisols are characterized by well-developed profiles with rich, dark-colored surface horizons. They form under grass.

Monsoon   Seasonal land and sea winds caused by the reversal of atmospheric pressure from one season to another over adjacent land and water areas.

Moraine   An accumulation of unsorted rock materials deposited during glacial action.

Mudflow   A mass of fine-grained earth materials with high water content that moves downslope as a fluid mass.

Muskeg   Waterlogged depressions in subarctic regions, filled largely with low-growing vegetation.

Natural levee   A long, narrow embankment formed by the alluvium deposited by a river where it overflows its banks.

Neck   A mass of solidified lava that fills the vent of a volcano.

Neutral shoreline   A coastline along which there has been no appreciable change in elevation so that deposition is the most active process.
Nonclastic sedimentary rock   A chemically or organically precipitated sedimentary rock; for example, limestone.

Occlusion   The overtaking of a warm front by a cold front in a low-pressure cell.
Orographic precipitation   Precipitation triggered by the lifting of masses of air due to their collision with a large surface feature, for example, a mountain or an elevated coastline.
Outwash   Clay, sand, or gravel carried by meltwater streams flowing from glaciers.
Oxbow lake   A semicircular body of water formed in the abandoned meander of a stream.
Oxidation   A form of chemical weathering that involves the combination of oxygen with other elements.
Oxisol   One of the ten orders of the U.S. Comprehensive Soil Classification. Oxisols are characterized by intense leaching that results in an oxic horizon. They are common in hot, humid regions.

Pahoehoe lava   A solidified lava flow that has smooth, rounded features.
Pangaea   The name given to the single large landmass on the earth which 200 million years ago began to split into the present continents.
Parallel   A line on a map drawn parallel to the equator.
Parent material   The unconsolidated and more or less chemically weathered rock or organic matter from which soils develop.
Paternoster lake   One of a string or a series of lakes in a glaciated valley; separated by rock steps.
Pedestal rocks   Isolated masses of rock above and resting on a smaller base that has been abraded by wind.
Pediment   A gently inclined bedrock erosion surface along the base of many mountain fronts in arid and semiarid regions.
Pediplain   Widely extending, gentle erosion surface formed by the coalescence of many pediments.
Peneplain   An old-age erosional surface where all landform features are reduced to a flat plain.
Perihelion   The place in the orbit of a planet where the planet is closest to the sun. The earth's perihelion occurs on January 3.
Permafrost   The zone of permanently frozen ground below the surface.
pH   A quantitative scale indicating the degree of soil acidity or alkalinity; measured by the negative logarithm of the hydrogen-ion activity of a soil.
Photosynthesis   The food-making process in plants.
Plant community   A distinct group of plants whose members have about the same ecological requirements.
Plant life-forms   Categories of plants based on the form into which they develop; trees, shrubs, herbs (including grasses), and mosses.
Plant succession   The sequence of plant communities that evolves on a bare land surface and proceeds toward the climax stage.
Plateau   An elevated flat plain that is limited on one or more sides by a steep slope.
Plate tectonics   A theory regarding the movement of the lithospheric plates of the earth's crust.
Playa   A shallow, enclosed desert basin that collects water following a rain.

Pleistocene epoch The period in geologic time during which glaciers advanced and retreated over large sections of the earth's surface.

Plucking A form of glacial erosion by which rocks are frozen to a moving glacier and are pulled away from their parent masses.

Plutonic rock Intrusive igneous rock that has cooled slowly at considerable depth in the earth's crust.

Podzolization A soil-forming process common in cool to temperate, humid climates. Podzols are characterized by a highly leached A_2 horizon.

Polar front A discontinuous front which divides polar and tropical air masses and along which most of the traveling cyclonic disturbances of the middle latitudes form.

Porphyry An igneous rock that contains large crystals set in a mass of finer-textured particles.

Prairie grass A continuous cover of grasses 1½ to 2 feet high that occupies temperate latitudes where annual moisture averages 15 to 25 inches.

Pressure gradient force The force caused by the rate of change of pressure over distance.

Pyroclastic rocks Rocks formed from the consolidated and compacted fragments of volcanic material; that is, lava, cinders, and ash.

Rejuvenation A change in landforms such that erosion activity is renewed.

Relative humidity The ratio (expressed as a percentage) between the amount of water vapor contained in a given volume of air and the total amount of water vapor that the volume of air can hold when saturated at the same temperature.

Revolution The movement of the earth around the sun. One revolution takes 365¼ days.

Ria shoreline A coastline typified by funnel-shaped land indentations made by the sea; caused by submergence.

Rock A solid, natural aggregation of one or more minerals.

Rockfall A free fall of a rock down a steep mountain slope.

Rockslide A mass of rocks that slide as a unit downslope under the influence of gravity.

Rotation The motion of the earth as it turns on its axis.

Saltation The process by which solid particles move in a series of hops along the bed of a stream or across a desert surface.

Sand dune A hill of sand constructed by wind deposition.

Saturation The state of the atmosphere when it can hold no more water vapor.

Saturation zone The subsurface below the upper limit of the water table.

Savanna Open tropical grassland with grasses of varying length and density, and scattered trees and bushes.

Sclerophyll A woody xerophytic plant with shiny, hard-surfaced, narrow, and often waxy, leaves.

Scoured rock Rock surfaces that have been grooved and polished by glacial debris dragged by a moving glacier.

Scrub A vegetation association characterized by stunted trees and bushes.

Sea-floor spreading The enlargement of the gap between two diverging lithospheric plates in an ocean region.

Sea stack A pillar of rock found as an erosional remnant along rocky coastlines.

Sedimentary rock A rock formed by the consolidation of rock fragments or sediment.

Seismograph An instrument that records and graphs ground tremors, especially earthquakes.

Selva Tropical rainforest vegetation composed of broad-leaved evergreen trees.

Sial The surface rocks of the continental crust, composed largely of Silica and Aluminum.

Sill A tabular pluton that lies parallel to the surrounding rock layers.

Sima   The dense rocks underlying continental landmasses and forming the ocean floor; composed largely of Silica and Magnesium.

Slip face   The sheltered leeward side of a sand dune, steeper than the windward side.

Slip-off slope   A low, gently sloping deposition surface projecting from the opposite side of an undercut bank along the meander of a river.

Slump   A large mass of earth materials that moves downhill with a rotary motion.

Snowfield   An area where snow has endured and accumulated from year to year.

Soil   A natural complex of organic and mineral matter suited for plant growth.

Soil consistency   The resistance of soil to breaking apart.

Soil horizon   A horizontal layer in the soil profile that differs from adjacent layers in physical, chemical, or biological characteristics.

Soil permeability   The rate at which water, gas, and plant roots penetrate a soil.

Soil porosity   The proportion of space in a soil not occupied by solid particles.

Soil profile   A vertical cross-section of a soil through all its horizons and extending into the parent material.

Soil structure   The distinct pattern formed by the manner in which soil particles combine to form secondary particles.

Soil texture   The relative proportion of sand, silt, and clay particles in a soil.

Solar constant   The rate per unit area at which solar radiation is received at the outer limit of the atmosphere; 1.95 calories/cm^2/minute.

Solar radiation   Radiant energy transmitted from the sun.

Solifluction   The mass movement of saturated earth materials over a permafrost layer.

Solstice   The day of the furthest poleward position of the sun during the year.

Solum   The A and B horizons of the soil; also called the *growth zone.*

Solution   A form of stream transport in which particles are dissolved in water.

Source region   The area over which an air mass lies and from which it derives its characteristics.

Specific humidity   The weight of water vapor per unit weight of air; expressed in grams/kilogram.

Spodosol   One of the ten orders of the U.S. Comprehensive Soil Classification. Spodosols are characterized by a highly leached A_2 horizon. They are common in the higher latitudes.

Squall   Strong, gusty winds accompanied by heavy rains associated with a thunderstorm.

Stable air   Air that resists vertical motion.

Steppe   A mid-latitude grassland characterized by short bunch grass that is 4 to 8 inches in height and that forms a discontinuous vegetation cover in semiarid climate regions.

Stock   A massive pluton similar to a batholith but less than 40 square miles in extent.

Storm surge   A rapid rise of the sea above expected tidal heights during which water rushes inland; often associated with a hurricane.

Storm track   Pattern described by a moving cyclone over a horizontal distance.

Stratosphere   The layer of the atmosphere above the tropopause and below the mesosphere.

Stream capacity   The maximum amount of material that a stream can transport.

Stream competency   A measure of the ability of a stream to move particles of a certain size.

Stream discharge   The amount of water in a stream flowing past a given point in a given unit of time.

Stream gradient   The rate of change in elevation between the stream head and the stream mouth.

Stream load   The amount of material transported by a stream at a given moment.

Stream mouth   The point at which the stream terminates its flow.

Stream terrace   A relatively flat plain bordering a river; represents the remains of a former floodplain, at a higher level, dissected by renewed downcutting.

Stream velocity   The distance that water in a stream travels in a given unit of time.

Stream volume   The amount of water in a stream channel.

Striations   Grooves and scratches in bedrock caused by rock particles dragged across the surface by moving glaciers.

Subduction   The process by which one crustal plate sinks below another.

Sublimation   The change of state of a solid to a gas without passing through the liquid state, or vice versa; for example, ice changing directly to water vapor, or water vapor to ice.

Submerged coastline   A coastline that formed due to the sinking of the land relative to sea level.

Succulents   Plants that have the ability to store large volumes of water in their tissues; for example, cactus.

Suspension   A form of stream transport in which stream current keeps sediment suspended above the stream bed.

Syncline   A downfold in rock strata in which the layers bend upward from the bottom or axis of the fold.

Taiga   The high-latitude coniferous forest in North America.

Talus   The loose rock fragments that accumulate at the base of cliffs.

Tarn   A small lake formed when the basin of a cirque fills with water.

Tectonic   Pertaining to crustal rock strata that have been folded or faulted or subjected to vulcanism.

Thermosphere   The outermost layer of atmosphere that lies above the mesopause and is transitional to outer space.

Thunder   The sound given off by shock waves in the atmosphere; associated with a lightning stroke.

Till   An accumulation of unstratified glacial debris of all sizes.

Timberline   The elevation on mountains above which trees do not grow.

Tombolo   A bar that connects an island to the mainland.

Topography   The description of the surface features of an area.

Tornado   A whirling storm formed around an intense, low-pressure system; commonly associated with a rotating thunderstorm.

Transform plate motion   The horizontal shift of crustal plates.

Transpiration   Evaporation of water from the leaves of plants.

Tributary   A stream channel that connects with and feeds into a larger channel.

Troposphere   The layer of the atmosphere closest to the earth's surface.

Tundra   The treeless expanses of subarctic climate in North America and Eurasia.

Ultisol   One of the ten orders of the U.S. Comprehensive Soil Classification. Ultisols are characterized by reddish yellow surface horizons caused by the removal of silica by leaching.

Undercut slope   A steep slope along a river bank formed by the erosion of the bank by the stream along the outside bend of a meander.

Underfit stream   A stream that flows in a channel much larger than the present stream could possibly erode.

Undertow   The undercurrent resulting from the backflow of a breaking wave on a beach.

Unstable air   Air that rises vertically in the atmosphere.

Vapor pressure   The amount of pressure exerted in the air by water vapor.

Vent   The central opening in the earth's surface through which volcanic material is forced during an eruption.

Ventifact   A boulder, stone, or pebble that has been worn and polished on one side by wind action.

Vertical zonation   The development of plant life-forms at different heights above the earth's surface.

Vertisol   One of the ten orders of the U.S. Comprehensive Soil Classification. Vertisols are characterized by large amounts of clay which expand and contract, causing the soil to overturn.

Vulcanism   The process by which molten rock is forced into openings in the crust or onto the surface.

Wadi   A steep-sided ravine in a desert or semiarid region which contains water only when it is raining.

Warm front   The boundary between a region of advancing warm air and cooler air.

Waterspout   An intense, but small-scale, whirling low-pressure system that is similar to a tornado but that forms over water instead of land.

Water table   The upper limit of the part of the soil or permeable rock strata saturated with water.

Water vapor   Water in the gaseous state.

Wave-cut notch   A deep, narrow gouge in the base of a sea cliff.

Weather   The state of the atmosphere at any given time.

Weathering   The physical and chemical breakdown of rock material on the earth's surface caused by the elements of weather.

Wind chill   The lowering of sensible temperatures due to wind velocity.

Xerophyte   A type of plant that has special adaptations that protect it against the lack of moisture.

Yazoo tributary   A stream channel that flows a long distance behind a natural levee before it joins the main channel.

Index

References to book pages are given in lightface *type, followed by references to MEDIAPAK frame numbers, indicated in* **boldface.** *A MEDIAPAK reference such as* **"M17:F4–8"** *means that the index entry can be found in MEDIAPAK 17, Frames 4 through 8.*

Abrasion
   glacial, 285. **M18:F8**
   stream, 265. **M17:F9**
   wind, 304. **M19:F5**
Absorption, 23–24. **M2:F11**
Adiabatic lapse rates, 68–69.
   M4:F24–27
Advection, 43. **M3:F22**
Agazziz (Lake), 295. **M18:F38**
Air masses, 77. **M5:F1**
   characteristics, 78. **M5:F2–5**
   classification, 78–79. **M5:F6–8**
   North American types, 80. **M5:F9–15**
Albedo, 25. **M2:F12,14**
Alfisol, 164, 167, 168. **M10:F23–27**
Alluvial fan, 267. **M17:F17**
Alpine glaciers. *See* Glaciers
Anticyclones, 46, 81. **M3:F39,40,41;**
   M5:F20,40,41
Aphelion, 7, 8
Aquifers, 113. **M7:F5–10**
Arête, 288. **M18:F17**
Aridisol, 166. **M10:F9–12**
Artesian well, 113. **M7:F9**
Asthenosphere, 217. **M14:F5,7**

Avalanche. *See* Mass movement
Azimuthality, 319, 320

Badlands, 308. **M19:F20**
Barometers, 35, 36, 40. **M3:F4,5**
Base level, 268, 269. **M17:F18,19**
Batholith, 235, 236. **M15:F17**
Bergeron process, 63. **M4:F30**
Boreal forest. *See* Coniferous forest, high-
   latitude
Boulder train, 293, 294. **M18:F34**
Boyle, Robert, 36, 41. **M3:F11**
Broad-leaved evergreens, 186, 187. **M12:F3**
Butte, 308. **M19:F22**

Calcification, 157. **M9:F31**
Caldera, 233. **M15:F8–10**
Capillary action, 152. **M9:F15,31**
Capillary water. *See* Soil, water
Cementation. *See* Lithification
Chapparal, 189. **M12:F16**
Charles, Jacques, 36, 41. **M3:F11**
Chemical elements
   rocks, 203. **M13:F2**
   soils, 151. **M9:F11,12**

Chemical weathering, 250, 251. **M16: F7–10,13**
    carbonization, 250. **M16:F8**
    hydrolysis, 251. **M16:F9**
    oxidation, 250. **M16:F7**
Chinook. *See* Foehn winds
Circle of illumination, 9. **M1:F17,21,22**
Cirque, 287. **M18:F15,16**
Climax vegetation, 175. **M11:F2–4**
Climograph
    climate types, 140–42
    description, 125, 126. **M8:F2**
Clouds, 67, 68
Cloud seeding. *See* Bergeron process
Coasts, 308–15. **M19:F26–33**
Col, 288. **M18:F17**
Collision-coalescence process, 63. **M4:F31**
Compaction. *See* Lithification
Comprehensive Soil Classification System,
    160. **M10:F1.** *See also* Entisols;
    Vertisols; Inceptisols; Aridisols;
    Mollisols; Spodosols; Alfisols;
    Ultisols; Oxisols; Histosols
Condensation, 59. **M4:F9,23,24**
    latent heat of, 59. **M4:F9**
    nuclei, 62. **M4:F22,28**
Conduction, 21. **M2:F21**
Conformality, 319
Coniferous forests
    high-latitude, 193. **M12:F26–28**
    middle-latitude, 190. **M12:F24,25**
    subtropical, 189, 190. **M12:F17–19**
Continental drift, 211, 212. **M14:F6**
Continental glaciers. *See* Glaciers
Continental shelves, 309. **M19:F23,24**
Convection
    heat transfer, 26. **M2:F22**
    precipitation, 65. **M4:F36,37**
Cooling degree day, 30, 31
Coordinate, 6. **M1:F7**
Coriolis force, 44, 45. **M3:F28–34**
Creep. *See* Mass movements
Crystallization, 205. **M13:F6–11**
Cyclones, 46. **M3:F39–41**
    middle-latitude, 83, 87–89. **M5:F20, 24–28,36–41**

Davis, William Morris, 274. **M17:F32**
Deciduous forest, 190. **M12:F20,21**

Deflation, 303. **M19:F1–3**
Degradation, 245
Delta, 267. **M17:F16**
Desert
    climate, 131, 132. **M8:F17,18,21,25**
    vegetation, 194, 195. **M12:F36,37**
Desert pavement (Reg), 303. **M19:F4**
Dessication. *See* Lithification
Dew point, 62. **M4:F21,23,24**
Diastrophism, 237–41. **M15:F21,22.** *See
    also* Faults; Folding
Dike, 235, 236. **M15:F19**
Discharge, 263. **M17:F6,7**
Dokuchaiev, V., 160
Drift, 285, 286. **M18:F9,10**
Drumlin, 292. **M18:F31–33**
Dust storm, 304. **M19:F6**

Earthflow. *See* Mass movements
Earthquake, 215, 241–43. **M14:F1**
Earth's interior. *See also* Asthenosphere;
    Lithosphere; Moho
    crust, 215. **M14:F3,4**
    inner core, 215. **M14:F2**
    mantle, 215–17. **M14:F3**
    outer core, 215. **M14:F2**
Ecliptic, 8. **M1:F16**
Eluviation, 152, 167. **M9:F15,29,30**
Emerged shoreline, 314, 315. **M19:F35**
Entisol, 164. **M10:F2,3,8**
Ephemeral stream, 307. **M19:F15**
Epiphyte, 186. **M12:F4**
Equinox, 10. **M1:F22**
Equivalency (Equal areas), 318, 319
Erratic, 286. **M18:F11,34**
Esker, 294. **M18:F35**
Evaporation, 59. **M4:F4,7,12**
Exfoliation. *See* Mechanical weathering

Faults, 237–41. **M14:F15; M15:F27–29,34.**
    See also Graben; Horst
    normal, 238. **M15:F30**
    reverse, 239. **M15:F31**
    strike-slip, 239. **M15:F32**
Firn. *See* Glaciers, formation
Floodplain, 270, 271. **M17:F24,25,27,28**
Foehn winds, 49. **M3:F51**
Folding, 237. **M15:F23–26**
Friction, 45. **M3:F35–38**

Fronts, 65, 66, 79, 83, 85, 87. **M4:F39; M5: F16,30.** *See also* Polar front
  cold, 83, 85. **M5:F25,26,29,33,34**
  occluded, 83, 87. **M5:F28,35**
  warm, 85. **M5:F25,26,29,31,32**

Galileo, 35. **M3:F4**
Gay-Lussac, Joseph, 36. **M3:F11**
Geographic Grid System, 6, 7, 318, 323. **M1: F7–15**
Geoid, 2
Geomorphic cycle, 274–76. **M17:F32–35**
Geostrophic wind, 45. **M3:F32**
Geyser, 115. **M7:F14**
Glaciers
  alpine, 287–89. **M18:F14–24**
  continental, 291–95. **M18:F14,25–36**
  erosion, 285. **M18:F7,8**
  flow, 284. **M18:F5,6,21**
  formation, 283, 284. **M18:F2–4**
  glacio-fluvial, 295, 296. **M18:F37–41**
  materials, 285–86. **M18:F9–11,37**
Graben, 239. **M15:F33**
Gravitational water. *See* Soil, water
Great circle, 319
Greenhouse effect, 26. **M2:F20**
Groundwater
  aeration zone, 113. **M7:F5**
  defined, 112. **M7:F4**
  for irrigation, 115. **M7:F16**
  saturation zone, 113. **M7:F5**
Gust front. *See* Squall line

Hail, 65. **M4:F33–34**
Hanging valley, 288. **M18:F18**
Heating degree days, 30
Highland climate, 138, 139. **M8:F54–58**
Histosols, 69, 70. **M10:F34–36**
Horizons. *See* Soil, profile
Horn, 288. **M18:F17**
Horst, 239. **M15:F33**
Humid continental climate, 135, 136. **M8: F40,42**
Humidity, 58. **M4:F1**
  absolute, 60. **M4:F15**
  relative, 60. **M4:F17**
  specific, 60. **M4:F16**
Humid subtropical climate, 133. **M8: F31–33**

Humus, 156. **M9:F19**
Hurricane, 102–94. **M6:F23–28**
Hydrologic cycle (Water cycle), 59, 60. **M4: F10,11.** *See also* Water budget
Hydrophytes, 178. **M11:F10**
Hygroscopic nuclei, 62. **M4:F22**

Icecap climate, 137, 138. **M8:F47,51–53**
Igneous activity. *See* Vulcanism
Illuviation, 152. **M9:F15,29,30**
Inceptisol, 165, 166. **M10:F7,8**
Inclination, 8. **M1:F19**
Insolation, 25. **M2:F16,17**
International date line, 11–13
Intertropical convergence zone (ITCZ), 127, 128. **M8:F6,7,12,15**
Inversion, 29. **M2:F33–36**
Island arc. *See* Plate motions, convergent
Isobar, 43, 44. **M3:F8–10,27**
Isotherm, 30. **M2:F39,40**

Jet streams, 81–83. **M5:F19–23**
Jungle, 187. **M12:F8,10**

Kame, 295. **M18:F36**
Katabatic wind, 49. **M3:F50**
Kettle, 292, 295. **M18:F30**
Köppen, Wladimer, 120, 121. **M8:F1**

Laccolith, 237. **M15:F20**
Lacustrine plain, 295. **M18:F38,39**
Land breeze, 48. **M3:F47**
Lapse rate, 29. **M2:F32.** *See also* Adiabatic lapse rate
Latent heat
  of condensation, 59, 62, 68. **M4:F9**
  of vaporization, 58, 59. **M4:F8,23**
Laterization, 156. **M9:F28,29**
Latitude, 7. **M1:F10,11,14,15**
Lava flows, 234. **M15:F13,14**
Leaching, 152, 156, 167–69. **M9:F14–16, 29,30**
Lightning, 99, 100. **M6:F12–14**
Lithification, 206
  cementation, 206. **M13:F13**
  compaction, 206. **M13:F14**
  dessication, 206. **M13:F14**
Lithosphere, 217. **M14:F5,7**
Loam, 149. **M2:F9**

Loess, 306, 307. **M19:F13,14**
Longitude, 7. **M1:F13–15**
Longshore current, 311, 312. **M19:F28**
Long-wave radiation, 26. **M2:F18,19**

Magma, 204. **M13:F6**
Map, 317–29
   direction, 323–26
   essentials, 322–27
   properties, 318–22
   scale, 326–27
   topographic, 327–29
Maquis, 189. **M12:F16**
Marbut, C.F., 160
Marine west coast climate, 133–35. **M8: F34–38**
Mass movements
   avalanche, 255. **M16:F22**
   creep, 255. **M16:F24**
   defined, 252. **M16:F14,16**
   earthflows, 253. **M16:F20**
   mudflows, 253. **M16:F19**
   rapid (landslides), 253–55. **M16:F17–23**
   rockfall, 255. **M16:F23**
   rockslides, 253. **M16:F17,18**
   slow, 255, 256. **M16:F17,24,25**
   slumping, 253, 254. **M16:F21**
   solifluction, 256. **M16:F25**
Meander, 271, 272. **M17:F25,26**
Mechanical weathering, 249–50. **M16:F2–6**
   exfoliation, 250. **M16:F5**
   expansion and contraction, 249, 250. **M16:F3**
   frost action, 250. **M16:F4**
Mediterranean
   climate, 132, 133. **M8:F28–30**
   scrub forest, 188, 189. **M12,14–16**
Meridians, 6, 7, 321, 322. **M1:F8,9,12,13**
Mesa, 308. **M19:F22**
Mesophyte, 178. **M11:F10**
Mesothermal climates, 132–35. **M8:F27.** *See also* Mediterranean, climate; Humid subtropical climate; Marine west coast climate
Microthermal climates, 135–37. **M8:F39.** *See also* Humid continental climate; Subarctic climate
Mid-ocean ridge, 218. **M14:F10**
Millibar, 40. **M3:F5,6**

Mineral, 203. **M13:F3,4**
Mirage, 131. **M8:F26**
Mixed forest, 190. **M12:F22,23**
Moho, 215. **M14:F4**
Mohorovičić discontinuity. *See* Moho
Mollisol, 166, 167. **M10:F13–16**
Monsoon winds, 48. **M3:F48**
Moraines, 290. **M18:F22–24,28,29**
Mountain breeze, 48. **M3:F49**
Mudflow. *See* Mass movements
Muskeg, 193. **M12:F28**

Neutral shoreline, 315. **M19:F36**
Newton, Isaac, 43. **M3:F23,24**

Occlusion. *See* Fronts, occluded
Orographic process, 65. **M4:F38**
Outwash plain, 295. **M18:F37**
Oxisol, 169. **M10:F30–33**
Ozone, 16, 21. **M2:F11**

Pangaea, 221–24. **M14:F16**
Parallel, 6, 7, 321, 322. **M1:F8–11**
Parallelism, 8. **M1:F20**
Parent material, 151. **M9:F10,21**
Pediment, 307. **M19:F17**
Perihelion, 7, 8
Permafrost, 196. **M12:F39**
pH, 151. **M9:F12,13**
Photosynthesis, 178. **M11:F12**
Plant
   communities, 175. **M11:F1,2**
   life-forms, 175, 176. **M11:F5,7**
   succession, 175. **M11:F2**
Plate motion
   convergent, 218, 219. **M14:F11–14**
   divergent, 218. **M14:F9,10**
   historical, 221–24. **M14:F16–21**
   transform, 221. **M14:F15**
Plate tectonics, 217–21. **M14:F6.** *See also* Plate motions
Playa, 307. **M19:F18**
Pleistocene epoch, 286, 287. **M18:F12,13**
Plutons, 235. **M15:F16**
Podzolization, 156, 157. **M9:F30**
Polar easterlies, 48. **M3:F46**
Polar front, 79, 81, 83. **M5:F17,19,24,25**
Prairie grass, 193, 194. **M12:F34,35**

Precipitation, 60–67.   **M4:F35–44.** *See also*
Bergeron process; Collision-coalescence
process; Hail; Sleet; Snow
Pressure, 40–43.   **M3:F1**
related to temperature, 41.   **M3:F11–14**
vertical distribution, 40.   **M3:F3**
world distribution, 41–43.   **M3:F16–21**
Pressure gradient force, 43–45.   **M3:F25–27,**
31,32

Reflection. *See* Albedo
Revolution, 7, 8.   **M1:F16,18**
Rockfall. *See* Mass movements
Rock
cycle, 203–9.   **M13:F5,12,18,22–24**
igneous, 204, 205.   **M13:F6–11,17**
metamorphic, 208.   **M13:F18–21**
sedimentary, 206, 207.   **M13:F12–17**
Rockslides. *See* Mass movements
Rotation, 7, 11.   **M1:F17**

Sand dunes, 304–6.   **M19:F7–12**
barchan, 304.   **M19:F8**
longitudinal, 306.   **M19:F10**
parabolics, 305, 306.   **M19:F9**
transverse, 306.   **M19:F11**
Sand storm, 304.   **M19:F6**
Sea breeze, 48.   **M3:F47**
Sea-floor spreading, 218.   **M14:F9**
Seismograph, 215, 242.   **M14:F1,2**
Selective scattering, 23.   **M2:F10**
Selva. *See* Tropical rainforest, vegetation
Sial. *See* Earth's interior, crust
Sill, 236.   **M15:F19**
Sima. *See* Earth's interior, crust
Sleet, 65.   **M4:F33**
Slump. *See* Mass movements
Snow, 63.   **M4:F32**
Soil
chemical elements, 151.   **M9:F10,11**
color, 150.   **M9:F8**
consistency, 150.   **M9:F5**
defined, 145.   **M9:F1**
permeability, 150.   **M9:F6,7**
porosity, 150.   **M9:F6,7**
profile, 152.   **M9:F17–20**
soil-forming factors, 153–56.   **M9:F21–27**
structure, 150.   **M9:F4**
texture, 149.   **M9:F2,3**
water, 150, 151.   **M9:F9**

Solar constant, 23.   **M2:F9**
Solar radiation, 22.   **M2:F8,9**
Solifluction. *See* Mass movements
Solstice, 10.   **M1:F22**
Spodosol, 167.   **M10:F17–22**
Springs, 114.   **M7:F12,13**
Squall line, 97.   **M6:F8,9**
Steppe
climate, 130, 131.   **M8:F17–20**
vegetation, 193.   **M12:F32,33**
Storm tracks, 87.   **M5:F41**
Stream
deposition, 267.   **M17:F15–17**
erosion, 264, 265.   **M17:F2,7–10,21–35**
flow, 262–64.   **M17:F2–7**
transport, 265, 266.   **M17:F11–14**
Subarctic climate, 136, 137.   **M8:F43–46**
Subduction, 219, 221.   **M14:F12–14**
Submerged shoreline, 313, 314.   **M19:F34**

Taiga. *See* Coniferous forests, high-latitude
Talus. *See* Mass movements, rockfall
Tectonics, 228.   **M14:F6.** *See also*
Diastrophism; Plate tectonics; Vulcanism
Temperature, 28–33
scales, 32, 33
statistics, 30.   **M2:F27–29**
Thunder, 99, 100.   **M6:F14**
Thunderstorms, 95–100.   **M6:F2–14**
Tornadoes, 100–102.   **M6:F15–20**
Torricelli, 35, 36.   **M3:F4**
Trade winds. *See* Tropical easterlies
Transpiration, 59.   **M4:F4**
Tropical easterlies, 47, 48.   **M3:F43**
Tropical monsoon
climate, 128.   **M8:F8–10**
vegetation, 187.   **M12:F10,11**
Tropical rainforest
climate, 126–28.   **M8:F4–7**
vegetation, 186, 187.   **M12:F3–9**
Tropical savanna
climate, 128, 129.   **M8:F11–16**
vegetation, 193.   **M12:F30,31**
Tropical thorn forest, 188.   **M12:F12,13**
Tundra
climate, 137, 138.   **M8:F47–50**
vegetation, 196.   **M12:F38,39**
Turbulent heat exchange. *See* Conduction;
Convection
Typhoon. *See* Hurricane

Ultisol, 168, 169. **M10:F27–29**
Underfit stream, 295, 296. **M18:F40**

Valley breeze, 49. **M3:F49**
Vertical ray of the sun, 9, 10. **M1:F23–26**
Vertisol, 164, 165. **M10:F4–8**
Volcanoes, 231–34. **M15:F3–11**
  cinder cones, 233. **M15:F6**
  composite, 233. **M15:F5**
  shield, 231, 232. **M15:F4**
Vulcanism, 231–37
  extrusive, 231–33. **M15:F3–12**
  intrusive, 231, 234–37. **M15:F15–20**

Wadi, 307. **M19:F16**
Water budget (local), 116, 117. **M7:F1–3,
  20–29**
Waterspout, 102. **M6:F21**
Water table, 113. **M7:F5,6**
Water vapor, 58–60. **M4:F3,4,12–14,20**
Waves, 309–11. **M19:F25–27**
Wegener, Alfred, 211, 212
Westerlies, 48. **M3:F45**
Wind chill factor, 31, 32

Xerophyte, 176, 178. **M11:F10**